Environment as Capsule

All about Environment Education: Informally

The Author

Winner of Subhash Chandra Bose National Award for Excellence, an award recognized by UNESCO and Communal Harmony Award, Dr S K Nanda is an IAS officer (1978 batch) currently serving as Director in Government of India's Housing and Urban Development Corporation (HUDCO). At the field level he has worked as Collector in districts of Dangs, Baroda, Panchmahals, Banaskanta and Junagadh in Gujarat. Known as the Man of Management, Dr Nanda has been instrumental in managing various socio-political and natural crises in the state for last 39 years of his service as an administrator.

As Deputy Secretary, Health Department (1987-88) he worked as Programme Coordinator for newly launched mission on Universal Immunization. He has given his services for Department of Tourism (1983-84), Secretary, Information & Broadcasting (1997), Managing Director- Gujarat State Financial Corporation (1999-2001), Secretary- Youth Services & Culture (2001), Principal Secretary, Health Department (2001-2003), Principal Secretary, Food & Civil Supplies Department (2004-2007) and Principal Secretary, Forests & Environment Department (2007-2012), Additional Chief Secretary, Home Department of Gujarat Government (2012 to 2014) and CMD in Gujarat State Fertilizer Corporation Limited.

Dr S K Nanda has also served as Guardian Secretary of Dangs district, a remote and fully tribal pocket, for restoration of harmony and initiating development for peace and prosperity for 15 years.

Earlier he authored 4 books out which one happens to be his poetry collection called "Feelings", his other titles include "Tribal Migration" which is a reference guide for many researchers and subject enthusiasts and "Earthquake Reflections" which he wrote for National Institute of Disaster Management (NIDM), Government of India, New Delhi. His last book "Dangs: The little known paradise of India" revolves round the mystique and beautiful district Dangs in Gujarat.

Apart from all this, he has interest in vivid areas like Jagannath Cultural Centre at Gandhinagar where he is rendering skilled training, dance, yoga etc. other than proto type Puri temple for religious visitors. He mentors Ham Radio institute for Training in High frequency communication. He was instrumental in the formation of the first Tribal Ayurvedic Cooperative in India in Dangs District in early 80's. He has experience in handling all disasters at State level and National level like earthquake in Gujurat (2001), tsunami in South India (2005) etc and now working for Hindustan Scouts & Guides as National Chief Commissioner for grooming youth in social and youth matters. Counseling activities to train young minds and communication with various segments is a passion with him.

A Doctor of Philosophy (PhD) in Rural Economics, Dr Nanda holds degrees of LLM, LLB and Master of Arts (MA) in Political Science.

Environment as Capsule

All about Environment Education: Informally

Dr S K Nanda, *IAS*

2018

Daya Publishing House®

A Division of

Astral International Pvt. Ltd.

New Delhi – 110 002

Cataloging in Publication Data--DK
Courtesy: D.K. Agencies (P) Ltd. <docinfo@dkagencies.com>

Nanda, Sudip Kumar, author.
Environment as capsule : all about environment education: informally / Dr SK Nanda (IAS).
    pages cm
Includes bibliographical references.
ISBN 9789387057319 (HB_Int Edn)

    1. Environmental protection--India. 2. Sustainable development--India. 3. Environmental education--India.   I. Title.
LCC GE160.I4N36 2018 | DDC 363.700954   23

Resource & Critiquing	:	**Hardik Shah,** *IAS*
Editor	:	**Dr Deepak Acharya**
		Email: contact@deepakacharya.in
Concept Design & Pre-press by	:	**Shailesh Modi, Chetan Dave**
		Email: shaili_06@yahoo.co.in
		Photos & Illustrations: Internet (Under Creative Commons)
Published by	:	**Daya Publishing House®**
		A Division of
		Astral International Pvt Ltd
		– ISO 9001:2015 Certified Company –
		4736/23, Ansari Road, Darya Ganj
		New Delhi-110 002
		Ph. 011-43549197, 23278134
		E-mail: info@astralint.com
		Website: www.astralint.com
Printed at	:	**Replika Press Pvt Ltd**

Contents

Preface

I have had vivid memories in childhood of best landscapes in my countryside long coastline and fabulous forests with tree cover inhabited by wildlife. After three decades those landscapes had vanished, coastline spoilt and forest cover replaced by township. I am not holding any grudge against development to make humans happy nor I am utopian in my thought process. What I badly miss is the absence of harmony in nature in the process of growing up of generations. Generation one after another has to hold the key to maintain the balance to keep Nature smiling at us and we basking in the cradle of Nature to understand and give meaning to its manifestations.

After joining Civil service my interface with Dangs, where I lived for two years, made me discover and love Nature passionately to understand its impact and see beauty in every odd thing including set of beliefs, I noticed. Subsequently, my stint as a Guardian Secretary for two decades made me an astute believer in wilderness of Environment to understand the mature somber organized way of working of Nature which has encrypted its hidden essence in Environment for our Existence. This ranges from Farming to Fishing, Crop selection to Consumption, and Honey bees to Medicinal plants, all meant to redeem and remedy ills of human actions through benedictions of Nature and Environment.

Our love for Environment is always dependant on our understanding and knowledge not wisdom or intelligence. The latter has probably caused more havoc to redeem its message and meaning. This is the reason why I felt motivated to write a small hand book on Environment that surrounds us in this planet where we all come and go like in a stage. We need to reckon that elements like Wind or Air that brings motion, Earth or Soil which gives us space to footfall for our stay, Fire which ignites our life, Trees which maintain our life cycle and Sun and Moon which govern us every way, each minute and every corner are our heritage and we must hold their baton to understand them and see its relevance. Ancient texts and past civilizations have always harped on the essence of Nature than describing its beauty and engraved it on our lifestyle than making it an item for painting or drawing competitions.

Needless to say a painter or writer or artist can be creative only if he is influenced or moved. If no bird flies over sea or there is no vegetation on slopes of hills, the painter will always draw a naked or denuded hill and an empty blue space devoid of avian penetration. This is the role that Nature plays and its band for us unwittingly.

In the process of writing this book I fertilized my mind with thoughts of persons with whom I have worked. Hardik Shah is a person who has a deep understanding of what I believe and shared a lot of thought process and Deepak Acharya as a Scientist has given me more reasons to believe that we were all essentially right. Shailesh Modi, a graphic designer, had to do his job of translating my thoughts into motivating pictures or graphics for readers to relate Nature to their thought process. I owe a lot to them for the germination process in my mind.

I had the unique chance to serve as Forest and Environment Secretary in Gujarat where my obsession or fancy for Environment could find its deep roots. Expanding green Mangrove cover on coastline and creating space for more aquatic species to survive, imploring people to adopt farm forestry for augmenting incomes and enriching soil, relating trees to astronomy and good luck to sway them into cultural forestry, convincing Industrialists to do plantation to check carbon advance and absorb dust in air were some initial exercises in Environmental adventurism that really made the see change in ecological balance.

This goaded me further to script my thoughts form of a book that could get ingested in minds of common man and prepare their thought process to think anew afresh and remove all hindrances to the unification of Man with Nature. Hitherto all essays and scripts about Environment have been couched in technical verbose that has distanced thought process and made it uneasy for people to digest relate and invoke principles of Environment and it's logical ways and means .The science of Environment is no less truer than Mathematics is to calculations. The Sanatan Dharma principle namely "A creature exists alone not for himself but also beyond self" could get deeply rooted in this process of reforming our mind to work for reinforcing and renovating the Nature and to become its worthy custodian.

This book is dedicated to common man who need to go through the pages like a fairy tale reading and finally to discover Environment even if they do not remember a single word or remember a single example, that is the hidden purpose of writing this small cryptic and graphic picture filled book. Hope I serve Environment this way to save itself from throes of degradation.

Dr. S. K. Nanda

"Earth has enough to satisfy every
man's need,
but not every man's greed."

- Mahatma Gandhi

Preview

Religions across the world may have different views and beliefs on how to lead a good life, but one thing they all have in common is the underlying principle of **"Oneness of Life and Environment."** That is, one cannot exist without the other. Everything around us depends on the inseparable relationship that we have with our surroundings, fellow living beings and nature at large. We cannot change one without influencing a change on everything else.

According to Hinduism, "All living beings are sacred because they are parts of God, and should be treated with respect and compassion. This is because the soul can be reincarnated into any form of life."

According to Christian beliefs, "Our environment is one of the greatest examples we have of God's power. The word environment encompasses all of God's most beautiful and awesome works. The environment is his creation, a precious and holy resource with which he entrusted all humans the loving care and wise use of. God asked all humans to be stewards of the environment."

Buddhism emphasises that, "The effects of one's karma, both good and bad, manifest themselves in one's life and also in one's environment. Inevitably, in life, we will find ourselves in an environment which reflects our inner life state-whether that be our family, our workplace, our society, and so on."

And in Islam this concept is explained in many versus each concluding, "God has created everything in this universe in due proportion and measure both quantitatively and qualitatively."

Historically, humanities have functioned in collaboration with and felt a profound physical and spiritual connection with their natural environment. But as we became scientifically and technologically advanced, we have lost this most important touch with reality. For example, the creation of industrial cities at the cost of enormous tracts of land. This led to an ever-growing hunger to rule and abuse nature for our profit.

Human beings are the most intelligent species to walk the planet and yet, we have done more damage to our environment than all other species and natural forces put together. The way we are going, it won't be long before we will outdo the damage done so far.

We need to take some drastic steps to control anymore damage on an urgent basis. We owe our whole existence to environment and now is the time to understand everything there is to know about environment. We need to understand that our relationship with our environment is a two way street. We take care of our environment and environment takes care of us. A healthy environment is the most essential component for a healthy population and better quality of life. We spend so much time in our offices, classrooms and houses that we have almost forgotten to expose ourselves to the wonders of the nature and its importance in our everyday life.

The beauty of our planet is that despite all the waste from living beings and deaths of numerous species, our planet has the ability to clean itself. To the point that no waste or dead species is visible once it is cleansed off. There is a lot we need to know about environment.

There is a lot that we can do to make a sustainable living without exploiting the nature any further. Taking care of the environment is important for a healthy body and healthy life not just today but for future too.

With this book, I intend to share the information on environment and the tips that will come in handy to all of us to help make our planet cleaner and healthier place for us and our family and friends. Let's start with the basics.

1. What is Environment?

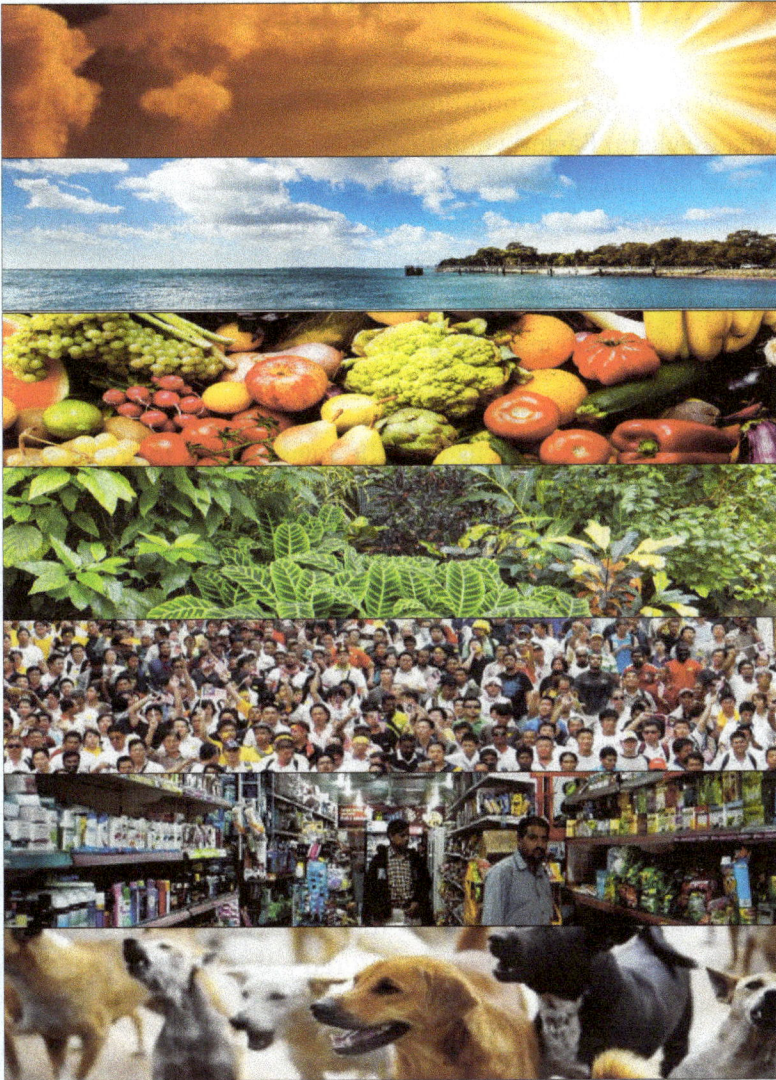

We think that the sky, the seas and oceans and plants and animals are all there is to environment. Well, environment is much more than that. Environment is all living things along with all the natural forces. Blowing wind, shining sun, people that we know, vegetables and fruits in our kitchens, barking dogs on the streets, a ride to school or office or super market; precisely everything that we come across is an integral part of our environment.

2. Significance of Environment

Plants take in carbon-dioxide and give out oxygen making the air we breathe clean. Bacteria and insects break down the organic matter to fertile the soil and provide nutrients for plants to grow. Birds and bees pollinate these plants so that they can continue to bear fruits and provide food for us thus fulfilling our nutritional requirements. There are millions of microbes unnoticed to eyes which make our soul and plant life and the richness of vegetation in its variety and phytochemicals they possess.

Significance of environment for human beings doesn't cease there. Environment also provides us with the raw material to create shelter for ourselves and our family, tools to make a living and medicines to heal our sick body. These natural medicines help cure numerous ailments including certain types of cancers. In USA, 40% of all the prescriptions are created from natural compounds from plants, microorganisms and animals. Taking care of environment is imperative for a healthy and happy life and future.

In the last 200 odd years, our relationship with our environment has taken a turn for the worse. The machines used in the factories release chemicals and gases as by-product and thus changing the aesthetics of the planet in every way. Many species of plants, animals and birds are becoming extinct across the world. The food we eat is laden with harmful chemicals (like mercury) and have entered our body. However, in the last 4 decades, more and more people have become aware of the situation at hand and realize that it is our responsibility to protect our environment, and our planet. Polluting environment will ultimately destroy us, our health and the beauty of this planet that is home to us.

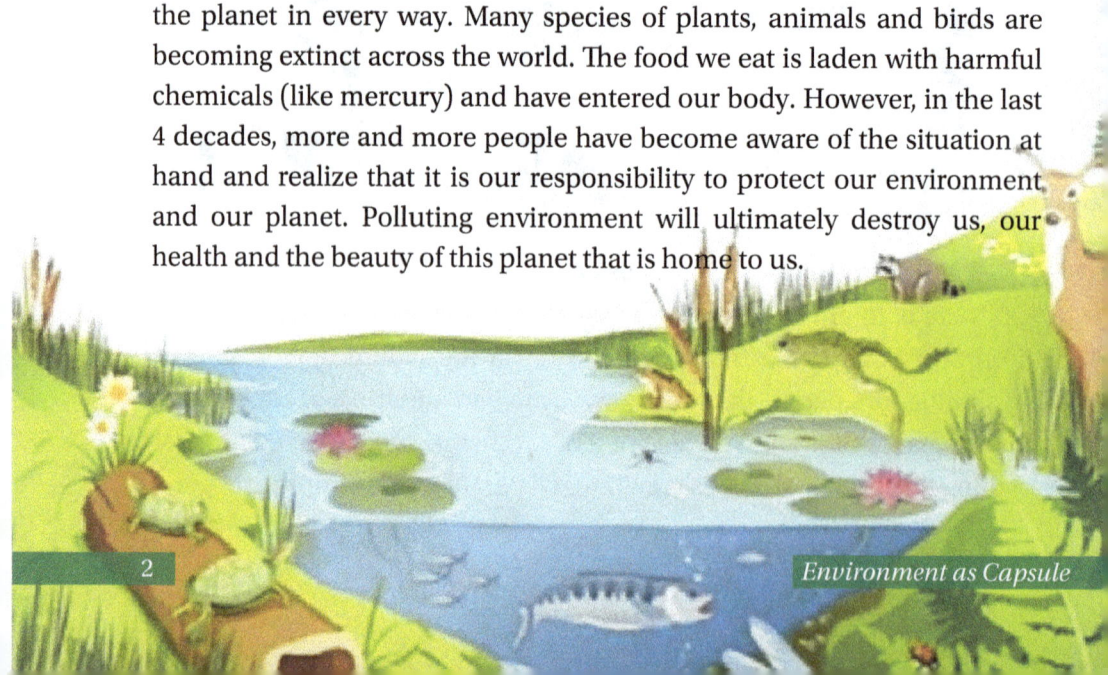

3. Three Thumb Rules of Preservation

- **Thumb Rule 1.** Universal Connection

- **Thumb Rule 2.** Every change in environment affects mankind

- **Thumb Rule 3.** We have a role to partake

Rule 1. Universal Connection

The first truth is that all things—seen and unseen—are connected. The concept of universal connection states that nothing in this world stands by itself. Every object is a link in an endless chain and is thus connected with all the other links. And this chain of the universe has never been broken; it unites all objects and processes in a single whole and thus has a universal character. We cannot move so much as our little finger without "disturbing" the whole universe. The life of the universe, its history lies in an infinite web of connections. Microbes in millions help in soil fertility and also act as bio control against pests. Agriculture is born out of this and sustaining livelihood and cattle leading to dairy for living.

Let's understand this with another example of honey bees. Honey bees pollinates crops while collecting nectar which allows them to bear fruit. But in the last decade, a large number of bees have died inexplicably in the Northern parts of India. Bees are no doubt small insect but their disappearance has adversely disturbed the web of connections in the life chain. Fruit bearing tress like apples, apricots, peaches, strawberries and many more are under a threat of extinction because they cannot bear fruits and reproduce without honey bees.

It may appear harmless to us right now. We may even think that we have just lesser fruits to eat. But, the truth is we have lost a very significant source of sustenance by unintentionally aiding the disappearance of honey bees.

Rule 2. Every Change in Environment Affects Mankind

Our health and the ability to survive depend entirely on the health of things around us. For thousands of years our species has adapted and survived on the planet and its challenging and ever changing environment. From the deep forests to Arctic tundra, man has raised above the natural challenges and lived. That said, toxic environment limits our ability to survive and thrive. We still need clean, healthy and fresh food to eat, water to drink and air to breathe. And we also need a shelter over our heads.

One of the biggest challenges our generation is facing is the phenomenon called Global Warming. Like all animals, humans also breathe out carbon- dioxide. Fortunately for us, plants absorb carbon-dioxide in order to grow, thus cleaning the air we breathe in. This also creates a natural cycle for carbon on the planet, globally. But in the past two centuries, human actions have seriously altered the carbon cycle. The smoke coming out from our vehicles, factories and farms is producing carbon-dioxide as well, thus causing an imbalance to the carbon cycle in the ecosystem. Cutting down forests has only accelerated this imbalance because the plants and trees that did not come under the axe, just can't keep up. Trees as source of carbon locks emit carbon once they are axed or used.

When plants can't absorb the excess carbon-dioxide, it is realized into the atmosphere where it floats (approximately 85 kilometers above us in the sky). It acts as a barricade, not allowing heat to escape the planet. This change has warmed up the planet by 1.4 degrees F in the last 150 years. It may look like a small figure, but many species (especially animals) can't adapt to this change so fast and thus become extinct. Scientists across the world are not sure about the actual implication of these changes in different regions, but one thing almost all of them agree to is that the climate will become more and more unpredictable and a lot more violent. Last year, Russia experienced its worst heat wave and drought in the history, 20% of Pakistan was flooded because of excessive rain and still experienced the hottest summer in the history.

That said, let's not lose hope...

Rule 3. We have a Role to Partake

Yes, that is right. We can all contribute to help preserve and protect our environment. Let›s see how:

- Conserving water
- Cleaning beaches
- Rain water harvesting
- Conserving energy
- Planting trees
- By starting recycling practices
- By not throwing trash/ garbage everywhere
- Using public transport to commute
- Donating things that are not useful to us instead of throwing
- them away

4. Lesser Known Facts

✓ Of 1.5 million known species, 16,118 species are endangered.

✓ Every year human beings consume 40% more resources than the nature can restore.

✓ An area of natural forest the size of Greece disappears every year.

✓ Human being across the world use about a million tons of paper every single day.

✓ We can save 24 trees for every ton of recycled paper we use, daily.

✓ Compact fluorescent light (CFLs) bulbs save 80% energy used by a standard bulb and last up to eight times longer. Latest is LED revolution which is more harmonious.

✓ Most of the world's creatures live in the sea. There are still millions of species that are yet to be discovered.

✓ On an average there are 27 oil spills per day in the world's oceans.

✓ Three out of four of the world's fisheries are fully or over fished.

✓ The Intergovernmental Panel on Climate Change predicts that the oceans will rise by 18-59 centimetres by the year 2100 because of melting of glaciers in Greenland and Antarctica. About 10% of the world's total population lives under a constant threat of floods.

✓ China, United States and India produce about half of the world's carbon dioxide emissions, Though India's share is just tolerable.

✓ China is building six enormous wind farms that will be able to produce 105,000 megawatts energy. Roughly the same capacity of France.

✓ Almost 1.1 billion people living in developing countries have insufficient access to water and 2.6 billion lack basic sanitation.

✓ The trajectories left by airplanes vapour trails make up almost half of the greenhouse warming caused by the airline industry.

✓ Almost all the plastic ever made still exists today. A plastic milk jug, for example, takes a million years to decompose. Only one to two percent of plastic used in the United States is recycled—the country produces 10.5 million tons a year.

✓ The World Health Organization estimates that 160,000 people die each year because of the indirect causes of climate change.

✓ More than 8,000 people die a day from breathing polluted air, largely due to inhaled coal particles which remain in their lungs.

✓ Carbon dioxide produced by burning fossils fuels is the primary greenhouse gas heating the earth. The levels today are higher than they have been for hundreds of thousands of years!

✓ Plankton are tiny creatures that serve as food for many sea animals. In sections of the Pacific there are six times more particles of plastic than plankton. Furthermore, worldwide levels of plankton have dropped 40% since the 1950s. The drop has been linked to rising ocean temperatures.

✓ The town of Chernobyl in the Ukraine is still one of the most toxic places to live because of a nuclear power plant meltdown in 1986.

✓ About 4 million pounds of trash is currently in space orbiting the earth. This is debris from satellites that have gone offline.

✓ Asia and sub-Saharan Africa are more prone droughts or floods and other extreme weather conditions linked to climate change. This will lead to shortage of food supply and deplorable living conditions.

✓ Fuels produced by plants like Sugarcane are a viable solution to petroleum-based fuels.

✓ Hooker Chemical disposed of about 22,000 tons of mixed chemical wastes into the Love Canal, a neighbourhood in Niagara Falls, New York, from 1942 – 1953. Love Canal was declared a threat to human health in 1978, and so the area had to be evacuated.

✓ Trees eat Carbon Dioxide and sea algae store carbon in its bed to make life easy for us to breathe fresh. Plantation of trees sucks carbon in air in its growth stage which provides oxygen to residents. These oxygen factories or green lungs also house birds to feed on bacteria and larva that cause epidemics in humans.

5. Sustainable Living

Sustainable living is a lifestyle that attempts to reduce an individual's or society's use of the Earth's natural resources and personal resources. Practitioners of sustainable living often attempt to reduce their carbon footprint by altering methods of transportation, energy consumption, and diet. Advocates of sustainable living aim to conduct their lives in ways that are consistent with sustainability, in natural balance and respectful of humanity's symbiotic relationship with the Earth's natural ecology and cycles .

Sustainable living is a buzz word these days. Though, I have to admit that not many people may know what it actually means. Essentially, sustainable living involves living as nonchalantly on the planet as possible which means that a person must leave as little impact on the environment as possible. Let us understand how: "Care for your needs but have concern and care for future days also"

In terms of accommodation, people who advocate sustainable living, build their homes in such a way that they use few nonrenewable resources. They do not require much energy to run, and cause little or no damage to the surrounding environment. Constructing a sustainable home requires materials produced in an environment friendly manner. For example, a home, which is made of hay rolls, or broken stone or bricks.

Sustainable living includes a dependence on sustainable energy. The energy sources used are renewable rather than limited in quantity. Instead of using non-renewable energy sources; such as fossil fuels, sustainable living uses renewable energy sources like solar, wind, water, or geothermal energy. To ensure that there is no damage to the environment, they capture and use the resources in an environment friendly manner. We need to download solar light into homes by providing roof reflectors that can conduit the light from sunrise to sunset in addition to conversion of solar heat to power through inverter.

The diet of someone who believes in sustainable living consists of foods that are at the base of the food chain. A vegetarian lifestyle is the greatest fit for sustainable living. It entails and ensures a minimalistic use of resources and guarantees production of edible food without causing environmental damage and degradation. It also requires crops, plants and trees growing organically because using insecticides pollute the environment and cause irreparable damage to flora and fauna. Local farmers markets are the best source for sustainably grown fruits and vegetables. Many people who condone sustainable living try to grow their own produce in their yard or in a community garden near their home.

6. Traditional Knowledge for Sustainability

The upsurge in the world's population and our ever-growing dependence on fossil fuel-based modes of production has played a considerable role in the growing concentration of greenhouse gases in the atmosphere. The global temperatures are increasing, the sea level is rising and precipitation patterns are changing. Storm outpourings, floods, droughts and heat waves are becoming more recurrent and severe. Subsequently, agricultural production is decreasing, fresh and clean water is becoming scarce, infectious diseases are escalating and source of revenue are unbecoming. Overall, the quality of human life and health is rapidly shrinking.

Therefore, traditional knowledge of sustainability of natural resources is exceedingly significant. To prevent the rapid depletion of natural resources, natural and social sciences are obtaining and applying all the information they have about our ecosystem, its conservation and restoration and are reinforcing the strategies and practices of sustainable development. Scientific research on human-environmental interactions is now a bourgeoning sustainability science. The theory accepts and recognizes that the well-being of human society depends largely upon the well-being of natural ecosystems. However, the intellectual resources realize that without contemplating upon the knowledge of native people, the foundation of sustainable science cannot be as strong as it should be.

Putting collective intellectual resources of both formal science and the knowledge that local folks have is the key to fostering a successful and acceptable sustainability theory and is important for the development we seek. Actually, people have even contended a Nobel Prize for sustainable living. Inspired by our current position, scientific research on human environmental interactions has developed into the new division of knowledge known as the Sustainability Science. Sustainability science seeks to comprehend the fundamental character of interactions between nature and society and collaborating universal processes with ecological and social features of specific places and sectors.

In fact, it will be suitable to propose that, science is not a monolithic entity; rather it is "a mosaic of the beliefs of many little scientific groups" with a diversity of viewpoints that individual scientists possess and then study objects deliberated to them. In numerous examples, scientists have only rediscovered what has always been a common knowledge locally. Therefore, saying that one is significant while the other is not, or one is more relevant than the other is not fair to either scientific research or to local knowledge. The differences that put them apart are: the way in which knowledge creation took place – and to some extent the way in the method of transmission – in both ways of knowing.

To postulate local knowledge as a non-science is boloney. However, that does not pledge an exclusive truth claim to either local knowledge or a blasphemy science. Any effort to inhibit information from re-examination and scrutiny, either by natives or by a curious researchers, endeavoring to learn a fresh process is not to challenge or discredit any specific system of knowledge.

Managing natural resources cannot be the subject of just any solitary form of knowledge, like the Western science. It has to consider a multitude of knowledge systems. There is a more essential purpose for integration of knowledge systems. Application of scientific research and local knowledge contributes to both- the equity, opportunity, security and empowerment of local communities, as well as to the sustainability of the natural resources. Local knowledge helps in situation analysis, data collection, management planning, creating the adaptive tactics to learn and acquire response, and institutional support to put these plans to practice. Science, on the other hand, provides advanced technology, or helps to improve existing technology. It also provides tools for networking, storing, visualizing, and analyzing information, as well as projecting lasting tendencies to help find efficient solutions to complex problems.

Local knowledge systems contribute to sustainability in diverse fields such as biodiversity conservation and maintenance of ecosystems services, tropical ecological and biocultural restoration, sustainable water management, genetic resource conservation and management of other natural resources. Moreover, they are also useful for ecosystem restoration and often have the elements of adaptive management.

7. Traditional Knowledge Systems for Biodiversity Conservation

Today the biggest challenge that stares humanity in the face is erosion of natural resources, from forests to water and agroecosystems across the globe. Not only are our natural resources at risk, households through farms, village and wilderness are under constant threat too. Traditional knowledge on biodiversity conservation in India is as diverse as 2753 communities and their geographical distribution, farming strategies, food habits, subsistence strategies, and cultural traditions. A multitude of environmental extortions created by unprecedented growth and consumerism have brought us to a point where one more mistake can lead to devastating results for all species that share this planet. Also imperiled is the sustainability of the essential ecological processes and life support systems in human dominated ecosystems across scales. Actually, human dominance of the planet is unmistakable in global change, biodiversity extinctions and ecological disruption.

Biodiversity conservation requires embracing specific local knowledge and institutional mechanisms. These are some of the most effective and valuable characteristics of local knowledge systems:

- Collaboration and collective action
- Intergenerational transmission of knowledge
- Skills and strategies
- Concern for well-being of future generations
- Belief in local resources
- Controlling resource exploitation
- Appreciation and respect for nature
- Proper management
- Conservation and sustainable use of biodiversity outside normal protected areas
- Transfer of useful species among the households, villages and larger landscape
- Future R & D in health sciences has to trail the plant chemical route than synthetic
- Chemicals for ensuring easy, cheap accessible and harmless drug formulation.

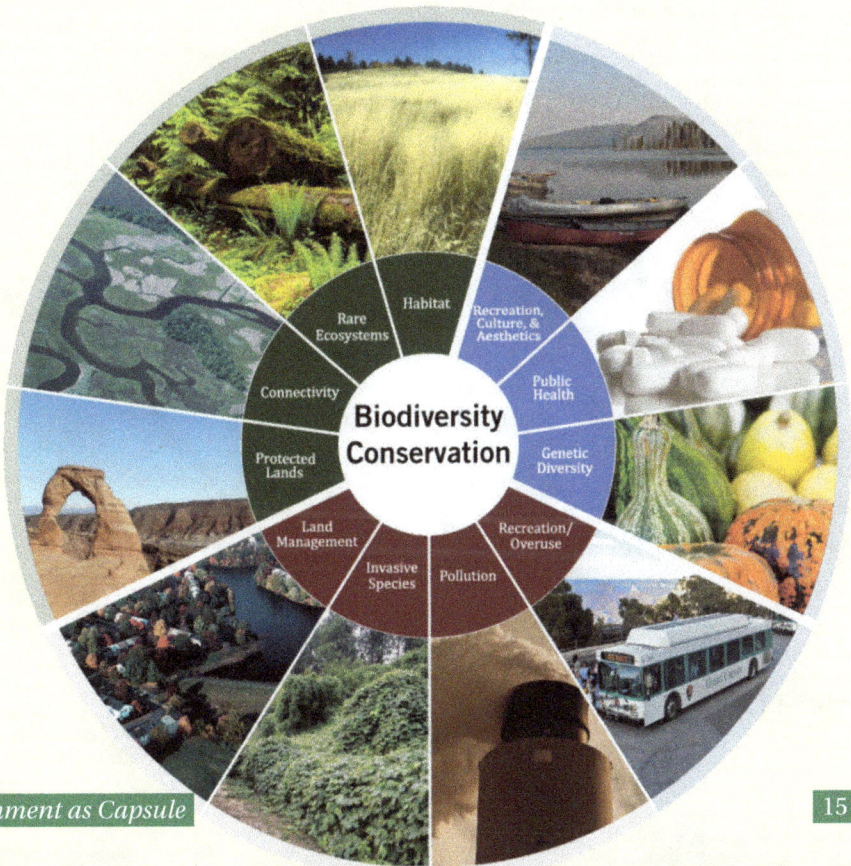

8. Local Vegetation Management

For a long time, local people have developed a range of vegetation management practices that continue to exist in tropical Asia, South America, Africa, and other parts of the world. People also follow ethics that often help them regulate interactions with their natural environment. Such systems require integration with traditional rainwater harvesting that promotes landscape heterogeneity through augmented growth of trees and other vegetation, which in turn support a variety of fauna. The practices are time tested, ecology proof and have signature of experience without need of an experiment to validate the science within its fold.

9. Conservation Principles in Ancient Texts

Ancient texts make explicit references to how forests and other natural resources are to be preserved and valued. Sustainability in various forms has been an issue of development of thought, since the ancient times. For example, robust principles designed to understand whether the intricate web of nature is sustaining itself or not. These principles roughly draw a parallel with modern understanding of conservation, utilization, and regeneration.

10. Conservation Principles

Atharva Veda (12.1.11) hymn, believed to have been composed sometime at around 800 BC, somewhere amidst deep forests reads, "O Earth! Pleasant be thy hills, snow-clad mountains and forests O numerous coloured, firm and protected Earth! On this earth I stand, undefeated, un-slain, unhurt." Implicit here are the following principles:

- It must be ensured that earth rested.

- It must be understood that humans can sustain only if earth is protected.

- To ensure that humans remain 'un-slain' and 'hurt' we must maintain the integrity of our ecosystem.

- Vaguely, it refers to ecology, economy and society concurrently.

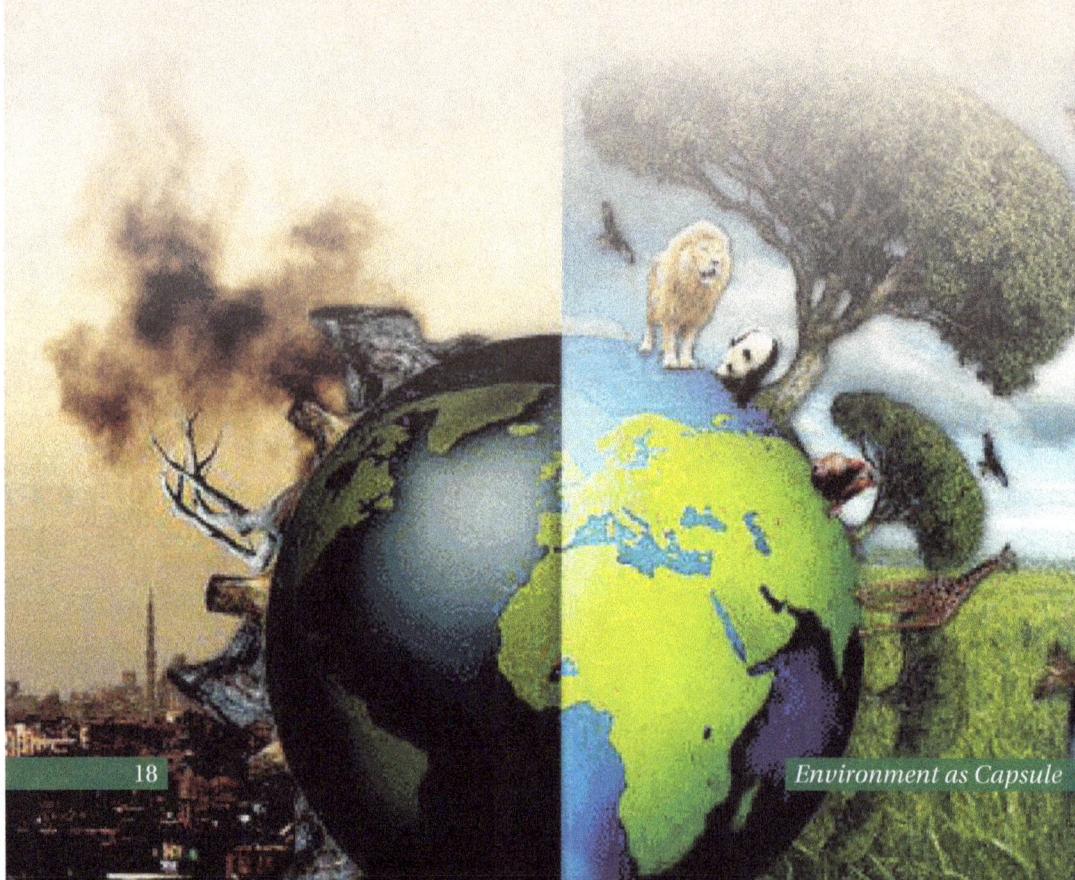

Environment as Capsule

11. Utilization and Regeneration Principles

Another hymn from Atharva Veda (12.1.35) reads, "Whatever I dig out from you, O Earth! May that have quick regeneration again; may we not damage thy vital habitat and heart". Implicit here are the following principles :

- Human beings can use the resources from the earth for their sustenance.

- Resource use pattern must also help in resource regeneration. In the process of harvest no damage should be done to the earth

- Humans are forewarned not against the use of nature for survival, but against the overuse and abuse.

Similarly, water management and associated tree growing has been the subject of ancient text. Tanks have been the most important source of irrigation in India. Some tanks may date as far back as the Rig Vedic period, around 1500 BC. The Rig Veda refers to lotus ponds (5.78.7), ponds that give life to frogs (7.103.2) and ponds of varying depths for bathing (10.71.7). Reference to the tanks is also found in the Arthashastra of Kautilya written around 300 BC. The Arthashastra refers to the ownership and management of the village tanks in the following verses:

- Ownership of waterworks such as reservoirs, embankments and tanks privately and the owner's wish to sell or mortgage them is considered appropriate (3.9.33.)

- The ownership of the tanks shall lapse, if they had not been used for a period of five years, excepting on case of distress (3.9.33.)

- Any one leasing, hiring, sharing or accepting a water works as a pledge with a right to use this, shall keep them in good condition (3.9.33.)

- Owners may give water to others in return for a share of the produce grown in the fields, parks or gardens (3.9.33.)

In the absence of owners, either charitable individuals or the people in village acting together shall maintain waterworks (3.10.3).

No one will sell or mortgage, directly or indirectly, a bund or embankment built and long used as a charitable public undertaking except when it is in ruins or has been abandoned (3.10.1.2).

The earliest scholar to have commented on the relationship of tanks and trees is Varahamihira who described the detailed technical instructions for the tank constructions in his famous work Brahatsamhita (550 AD). Without the shade of the trees on their sides, water reservoirs do not look charming; therefore, one ought to plant the gardens on the banks of the water (55.1)7.

Commenting on the species to be planted on the embankments of the tank, after its construction, Varahamihira writes: The shoreline (banks) of the tanks should be shaded (planted) with the mixed stands of Arjun (*Terminalia arjuna*), Vata (*Ficus benghalensis*), Aam (*Mangifera indica*), Pipal (*Ficus religiosa*), Nichul (*Nauclea orientalis*), Jambu (*Syzygium cuminii*), Vet (*Calamus*), Neep (*Mitragyna parvifolia*), Kurvak (*Larus ridibundus*), Tal (*Borassus flabelliform*), Ashok (*Saraca asoka*),Madhuk (*Madhuca indica*) and Bakul (*Mimusops elengi*) (54.119).

The substantial overlap in the formal and scientific forestry policy and practice, which provides hope that traditional knowledge systems can contribute to the management of natural resources, proves the need to understand and respect both. It would be pertinent to quote Gadgil and Guha (1992: 51) in this context, "Indeed one could argue that scientific prescriptions in industrial societies show little evidence of progress over the simple rule-of-thumb prescriptions for sustainable resource use and the conservation of diversity which characterized gatherer and peasant societies. Equally, the legal and codified procedures which are supposed to ensure the enforcement of scientific prescriptions work little better than earlier procedures based on religion or social convention."

12. Traditional Knowledge: Water Conservation

Societies, for thousands of years, have developed a diversity of local water harvesting and management regimes that continue to survive even today. There are many examples of this practice in South Asia, Africa, and other parts of the world. By integrating agroforestry and ethnoforestry practices, these systems continue to live on. The suggestions for sustainable water management, which looked promising in solving water conservation issues include taxing consumers to pay commensurate costs of supply, and distribution and of integrated watershed management and charging polluters for effluent treatment. Such measures are crucial, but they are insufficient and would need to draw on the local knowledge on rainwater harvesting across different cultures.

Rainwater harvesting in South Asia is different from other parts of the world in that it has a continued history of practice for at least over 5000 years. Similarly, Balinese water temple networks as complex adaptive systems are also very useful systems. Although hydraulic earthworks have occurred in ancient landscapes in many regions, they are no longer operational systems among the masses in the same proportion as in South Asia. For instance, we can find remains of earthworks and water storage adaptations, in Mayan lowlands in South America. Such systems were of use for prehistoric agriculture in Mayan lowlands, and for fish culture in Bolivian Amazon. Many find that rainwater harvesting is scientific and useful for rain fed areas. For instance, a validation comes from the Negev. Ancient stone mounds and water conduits found on hillslopes over large areas of the Negev desert. Field and laboratory studies suggest that ancient farmers were very efficient in harvesting water. A comparison of the volume of stones in the mounds to the volume of surface stones from the surrounding areas indicates that the ancient farmers removed only stones that had rested on the soil surface and left the embedded stones untouched. According to results of simulated rainfall experiments, this selective removal increased the volume of runoff generated over one square meter by almost 250% for small rainfall events compared to natural untreated soil surfaces.

One of the standard tree genus growing in association with tanks and ponds in India is Ficus, respected throughout the country, traditionally. It is a keystone genus, which supports anarray of other species. Records of frugivory from over 75 countries for 260 *Ficus* species (approximately 30% of described species), suggest that in addition to a small number of reptiles and fishes, 1274 bird and mammal species in 523 genera and 92 families are known to eat figs.

Traditional Knowledge Systems for Water Conservation

India is a huge country with a massive diversity not just in terms of her citizens, but also environmentally. Our landscapes and uncertain rainfall patterns, across all regions, make availability of water scarce for some while abundant for others. For years though, farmers have irrigated the farms with the quantity of water that is available to them by implanting and effectively imparting traditional knowledge of water conservation. Traditional knowledge of water conservation is strictly local and ancient. Effectively harvesting water by applying native approaches they have accomplished to use and share water with other species. Let us learn about a few traditional water conservation methods in India used by farmers and natives.

13. Katta

A provisional structure made by binding mud and loose stones available locally. Built across small streams and rivers, this stone bund slows the flow of water, and stores a large amount (depending upon its height) during the dry months. The collected water gradually seeps into ground and increase the water level of nearby wells. In coastal areas, they also minimize the flow of fresh water into the sea.

It is a cost effective and simple method, used widely in rural areas. Series of stone bunds built one behind the other have proved to be more effective than modern concrete dams in some villages, as these local structures can be easily repaired by farmers themselves. Although they require many skilled labourers during construction, all villagers contribute towards the cost of construction. However, with more people opting for personal bore wells and hand pumps, ground water depletion is occurring at an unmatchable speed, leaving the peripheral villages suffering. Thus, renewing and restoring these communal structures will play a vital role in sustainable life development. Even placing bags full of stones and sand in tracts of sluggish water movement can halt the water for few days to be harnessed for human advantage.

14. Sand Bores

Sand bores offer a safe alternative for farm irrigation without affecting groundwater. This technique uses the concept of extracting water retained by sand particles. Sand particles act as great water filters by absorbing and retaining the salt at the bottom of the sand bedwhile pumping clean water out; white sand has properties required to filter and make water drinkable. Sand deposits (as high as 15-30 feet) laying along the banks of the rivers are dug with a manual soil cutter and long PVC pipes are inserted into them, which act as filters and with the help of an electrical or diesel motor clean filtered water is gushed out.

This is another cost effective approach, which does not require a lot of maintenance. Karnataka is the leading example of the success of this technique, for decades. However, this technique is effective and successful only around coastline since the region has abundant sand deposits.

15. Johads

The water soak pits called as Madakas in Karnataka, Pemghara in Odisha and Johads in Rajasthan, and Tankis in the rest of the country, are one of the oldest and one of the most cost-effective systems of water conservation. These are mud and rubble barricades erected across the curve of a slope to detain rainwater, with a high embankment on the three sides and an open fourth side from where the water enters. These earthen dams catch and conserve rainwater, leading to improved percolation and groundwater recharge. Johads collect monsoon water, which slowly seeps in to recharge groundwater and maintain soil moisture. While many of these storage tanks are stand-alone constructions, sometimes it becomes necessary to interconnect them to cater to a larger group of flora and fauna of the region, to interconnect them deep channels with a single outlet in a river or a stream is constructed. This prevents structural damage.

Water from Johads is still been widely used by farmers to irrigate fields in many parts of India. Rajendra Singh of Tarun Bharat Sangh has put revolutionary efforts in reviving these communal water tanks and arid state of Rajasthan has seen a drastic improvement in water conservation. However, several other places that became dump yards for waste and places that fell prey to weeds can benefit from similar efforts put in by the locals.

16. Bawdi/Jhalara

These step-wellsare grand structures of high archaeological significance constructed since ancient times, mainly in honor of kings and queens. They are typically square shaped step-wells with beautiful arches, motifs and sometimes rooms on sides. Located away from residential areas, the water quality in these Bawdis is good for consumption. The typical lifespan of Jhalaras is around 20-30 years. Built with large investment of money and numerous skilled labourers, these magnificent structures today stand discarded by society.

Many of these structures have been encroached becoming mere dumping sites. Renovation of few of them in Rajasthan has brought their huge water storing capacity back, and use of electric pumps to draw water, these structures have saved lives during dry periods. Gujarat, Rajasthan and Karnataka have the maximum number of Bawdis, which attract tourists from all over the world.

17. Bamboo Drip Irrigation

Originated by tribes of North Eastern states, this technique economically uses water during dry seasons. This technique is especially proficient in hilly areas, since construction of ground channels is not possible in such regions due to sloppy and stony topography. This arrangement taps spring water to irrigate fields. A network of channels made by bamboo pipes of various diameters, which (to control the flow), allows downward flow of water by gravity. An efficient system can reduce around 20 liters of inflow water running over km to 20-80 drops per minute in agricultural fields.

Construction material such as bamboo and fiber is locally available. It is cost effective requiring less maintenance and only 1-2 labourers, who use tools to create a network of bamboo pipes to irrigate one hectare of land in 15 days. The system lasts for around three years after which the wood rots and decomposes to become nutrient-rich soil. Due to its effective and successful results, many urban regions have also replicated this model, where water is stored on roof top tanks and it flows through bamboo channels to irrigate fields and back gardens. Main advantage of the system is that it does not pollute like plastic counterparts and is very economical and simple to construct.

18. Integration of Traditional and Formal Science

There has been an ongoing debate whether integration of modern science with ethno science is imaginable. All the experimental and experiential evidence suggests that this integration is possible and will help mankind save the planet. In fact, traditional knowledge complements scientific knowledge by providing practical experience in living and responding to ecosystem change. However, the "language" of traditional ecology is different from the scientific and generally includes "figurative imagery and spiritual expression, signifying differences in context, motive, and conceptual underpinnings".

Indian traditions and local knowledge have repeatedly cemented the path for many discoveries in science. In fact, growth of science in India depended upon the knowledge and the wisdom created in ancient times on a range of disciplines including metallurgy, mathematics, medicine, surgery and natural resource management. Traditional skills, indigenous practices, and rusticabilities provide an extensive gamut of knowledge, and since "knowledge cannot be fragmented"; we have to consider the validated local knowledge apart from science for progressing a vigorous sustainability science.

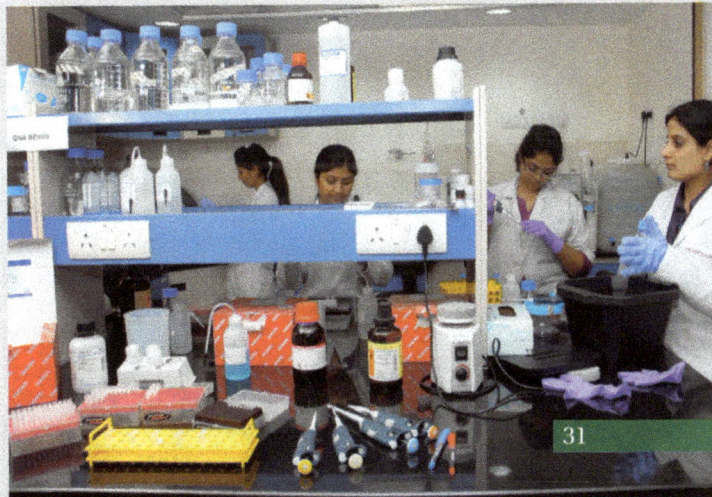

Environment as Capsule

The boundaries between modern and traditional systems of knowledge may be imaginary. Superficial limitations may just be the unfamiliar and unexplored province that challenges the cognizance for the need of interdisciplinary assessments. However, these confines are reducing and becoming convergent. Our scientists are progressively recognizing and accepting that, "there is a continuum between artificially dichotomized aspects of science: objective versus subjective, value free versus value laden, neutral versus advocacy." This corrective montage will have profound influence on science and policy development.

Since most masses in India are still implementing traditional and native knowledge systems, they can contribute to look into the trials of forest management, sustainable water management, biodiversity conservation, and mitigation of global climate change. The consequences of ecological imbalance demand that we get into stockpiles of knowledge and then form our mitigation strategies.

Any effort, striving to integrate traditional knowledge for biodiversity conservation and sustainability of natural resources must follow the principle that; "traditional knowledge often cannot be separated from its cultural and institutional setting." Few suggestions that may come in handy while integrating cultural and institutional knowledge are as follow:

- Each program aiming at the promotion of traditional knowledge needs acknowledgment that natural resource rights and tenure security of local communities forms the fundamental basis of respecting traditional knowledge.

- Additional consideration on fortification of intellectual property rights of indigenous people.

- Ground breaking projects, that warrant the augmentation of the capacity of local communities to use, express and develop their traditional knowledge based on their own cultural and institutional norms, may need priority while formulating policies.

- Making means for mutual learning between local people and the formal scientists available.

- State forest policies and sustainable forest management processes offering full attention to ethnoforestry and local institutional arrangements to incorporate traditional knowledge in forest management and development projects.

- Traditional knowledge and traditions can contribute to the preparation of village micro plans, which are prepared for ecodevelopment, joint forest management and rural development. The plans should be based on both geographic and traditional community boundaries rather than only on administrative boundaries.

- We have had water management systems serve the society for hundreds of years in our country. However, they stand susceptible today and need restoration.

- Sharing observations, findings and taking feedback to succeed in coming up with flexible and thorough strategies for management of natural resources.

- Despite understanding the importance of traditional knowledge for biodiversity conservation and natural resources management there still is a need to further the cause. The following consideration may be useful in this respect:

  - By encouraging proper documentation of indigenous knowledge and its use in natural resource management, prepared by the participating country that holds the knowledge. The records should not only contain accounts of knowledge system and its usage, but also data on issues that threaten our survival. Peoples biodiversity registers are an excellent example and starting point. PBRs encourage recording all traditional and conventional knowledge and acumen on use of natural resources that aid local educational institutions, teachers, students, government and NGOs that work as a team. By formulating ways and methods for maintenance, PRRs continuously make new and improved setting for sustained practice, which serves a substantial part in "upholding sustainable, flexible, and participatory system of management and in warranting a healthier course of welfares from economic use of living resources to the local communities."

  - By translating existing document into local languages and distributing them to locals. For example, narrating ancient texts on medicinal plants. Translation of local knowledge into formal scientific terminology offers space to external researchers, policy makers, and practitioners to understand and support people's knowledge systems and initiatives.

  - Facilitating the exchange of information amongst practitioners of local knowledge.

  - Developing clear and concise educational material on traditional knowledge systems and using it in communication programmes to convey information concerning the facts and fears to indigenous knowledge systems to both policy makers and the public.

  - Scientific institutions play an imperative part in supporting the knowledge systems. As has been pointed out earlier, contradictions between local and formal systems of knowledge is not real, and that any knowledge requires a fixed set of basic principles, theories and paradigms to build upon.

19. Environmental Sustainability Issues in India

India makes up 2.4 percent of the world's land and homes 16% of the world's population. The commingling outcome is a relentlessly unsustainable use of natural resources for several generations. Currently, India is experiencing rapid and widespread environmental degradation at alarming rates. Tremendous pressure is upon the country's land and natural resources to support the massive overpopulation.

Negligence and overuse of once plentiful forests has ensued in desertification, adulteration, and soil depletion throughout the subcontinent. This has grave corollaries for the livelihoods of hundreds of millions of Indians that live off the land. In Rajasthan alone, approximately five million tribal people depend on the collection of forest produce as their only source of income or nourishment. Without continual access to forest products such as fruit, honey, or firewood these communities experience debilitating hunger and are reduced to extreme poverty. Drought is devouring austere resources for the people of Rajasthan who endure chronic shortages of water. Onefifth of the villages in Rajasthan reported that they had no access to a reliable water source, and approximately half relied on a single source for the entire area. This affects the availability of safe drinking water, the success of the livestock population, and the security of basic food sources. Without adequate water, debilitating health and productivities take a toll of these people from this dry and arid state.

There is a bigger necessity for all concerned stake holders (government, NGOs, scientific institutions, and local boards) to conduct research on subjects such as soil stabilization, organic farming, erosion prevention, and protection and management of forested lands, often. Together they would search for practicable elucidations to environ mental problems and then offer the local community with essential backing and substructure. Aid from environmental organizations is an invaluable way for the deprived to mend the conditions of their local environment, directly affecting the quality of their livelihoods.

- In India, classification of these systems is based on various traditions which is as follows:

  - Religious traditions: temple forests, monastery forests, sanctified and deified trees

  - Traditional tribal traditions: sacred forests, sacred groves and sacred trees

  - Royal traditions: royal hunting preserves, elephant forests, royal gardens etc.

  - Livelihood traditions: forests and groves serving as cultural and social space and source of livelihood products and services.

Environment as Capsule

We can see the reflection of these traditions in a variety of practices concerning the usage and management of trees, forests and water. These include:

- Collection and management of wood and non-wood forest products.

- Traditional ethics, norms and practices for restrained use of forests, water and other natural resources.

- Traditional practices of protection, production and regeneration of forests.

- Cultivation of useful trees in cultural landscapes and agroforestry systems.

- Creation and maintenance of traditional water harvesting systems such as tanks along with plantation of the tree groves in the proximity.

These systems offer sustenance to biodiversity, which helps reduce the harvest pressure. Environmental ethics of Bisnoi community in Rajasthan suggests compassion to wildlife, and prohibits demolishing of Prosopis cineraria trees found in the region. Bisnoi teachings proclaim,"If one has to lose head (life) for saving a tree, know that the bargain is inexpensive."

Indigenous methods of managing vegetation originate from the fundamental ecological notions of native communities, in India. Two key characteristics of these systems are that the unit of nature, in terms of a geographical boundary; and a biotic components, plants, animals, and humans within this unit are considered interlinked. Many local knowledge systems offer parallel disposition to the evolving scientific view of ecosystems as erratic and irrepressible, and of ecosystem processes as nonlinear, and full of surprises.

We find contingents of authentic traditions by identifying animal and birds associated with various deities and use of various colours in flowers as offerings to various Gods and Goddesses. These meant propagation and protection floral/ fauna for sublime cause.

20. Biodiversity in Sacred Cliffs

Cliffs are downright overlooking cultural landscape with elements that upkeep a variety of species of plants and animals in India. As humans have special fascinations to cliffs,they consider them sacred. Cliffs support undamaged ancient woodland, dominated by tiny, slow-growing and widely spaced trees. Vertical cliffs conserve colossal range of widely spaced trees that are remarkably old, deformed and slow growing. On cliffs, we can find some of the most ancient and least-disturbed wooded habitats on Earth, even if such locations are close to concentrated agricultural and industrial development. The age of the trees on cliffs may indicate the age and growth rates of the entire plant groups there. Many examples of such habitats thrive in India.

When researchers surveyed 7 cliffs with ancient vegetation in Udaipur and Kota districts of Rajasthan, they observed that the cliffs were home to more than 25 species of trees, several species of shrubs and herbs. Regions close to Bhopal have more than 50 cliffs in central India in a radius of about 100 km. All the 7 cliffs surveyed in Rajasthan are sacred. They are part of the sacrosanct passages along the riverbank escarpment with more than a few meters of precipitous fall.

21. Farm Biodiversity

We can find domains of flora encompassing several species of plants and small animals, throughout the Indian farms and fields. These strips are valuable in many ways. Such strips on tropical lands accelerate natural successional processes by attracting seed-dispersing animals and increasing the seed rain of forest plants. Effects of these bands resemble the windbreaks on seed deposition patterns. Secluded trees provide seed for natural regeneration in the region. Because such strips trap large number of seeds of several species, they aid further tree growth. Parallel to open fields, farm boundaries with vegetation receive seeds in larger densities and species richness than open farms and pastures. Farm boundaries maintained throughout the country are often self regenerating and entail only management as these obstacles considerably increase the deposition of tree and shrub seeds within the cultural landscape. This practice needs to be maintained as it has several socio-economic benefits as well.

22. Cultivation of Medicinal Plants

Plentiful examples of medicinal plant cultivation by local people in India are inked in our ancient texts. Socio-culturally valued species find place in home gardens and courtyards. For example, Around the Nanda Devi Biosphere Reserve in the western Himalaya, the Bhotiya community, whose livelihood depends on local natural resources, practices seasonal and altitudinal migration and stays inside the buffer zone from May-October. Traditional knowledge of Kumaon Higher Himalaya establishes that Bhotiya clans use 34 species of medicinal plants native to the region. Among these, *Angelica glauca* and *Allium stracheyi* are narrow range endemic and *Allium stracheyi*, *Picrorhiza kurrooa* and *Nardostachys grandiflora* have place in the Red Data Book of Indian Plants. Interestingly, the annual production of medicinal plants is comparable with the annual production of traditional crops. Thus, cultivation, and harvesting can help securing their income and conservation of these species. Similarly, Juang and Munda tribes of the Keonjhar district of eastern India use 215 plants, belonging to 150 genera and 82 families. This advocates a wealth of traditional knowledge on biodiversity and herbal health care in tribes of eastern India. Tribes in the region are dependent on forests for other species as species that are included in their diet including cooking oil.

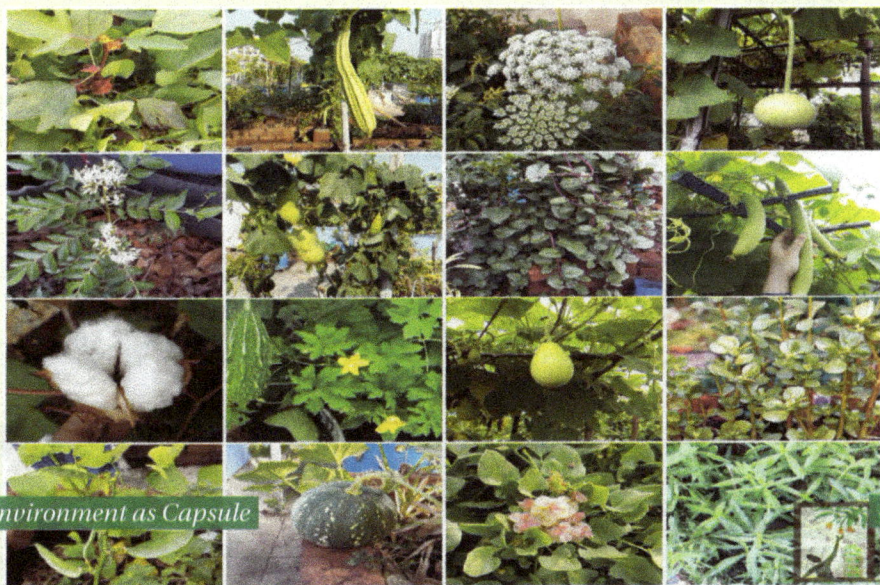

23. Traditional Ethos

Despite modernization, traditional ecological ethos continues to survive in many other local cultures, even if in abridged forms. When we investigate rural Bengal communities and their use of traditional norms and approaches, we find that a large number of biodiversity is safe and flourishing, regardless of their value and importance. These indigenous people may not know the concept of conservation and its importance for environmental protection, but their love, respect and consideration for nature and all life forms is evident.

Another example from northeast, which is particularly distinguished, is the tribal communities of Meghalaya – Khasis, Garos, and Jaintias – their religious beliefs that encourage conservation. In many other regions throughout the country, patches of forests have been titled sacred groves under customary law and these regions remain protected from any product extraction by the community. Such forests are very rich in biological diversity and harbor many endangered plant species including rare herbs and medicinal plants.

In Western Ghat region, the hills and cliffs which are treasure trove of rich bio diverse species of plants are named after, Pandavas, Hidamba, Sita Mata and Hanuman or Bhim to give honour to these regions that automatically bay for preservation theme.

Environment as Capsule

An old custom worth mentioning is use of plants in mural painting. For example, the Ajantan mural art. The practice covered a whole millennium from the second century B.C. to the eighth century A.D. The tradition continued up to the nineteenth century under the support of different dynasties in India, but declined by the end of that century.

Formal conservation efforts in India have relied profoundly on the recently affirmed official protected areas in various categories for biodiversity conservation. However, ancient and widespread human practice to set aside areas for the preservation of natural values in.

Keeping the rapid degradation of biological and cultural landscape and quality of life in mind, biodiversity conservation based on the "human-nature interactions," is extremely important to build sustainable life. We need environmental and cultural revolution, aiming at the reconciliation of human society with nature.

It is here that one has to allude to our current religious beliefs and traditions that have been solely embedded in conservation of species of flora and fauna- Lord Shiva with serpent of body, Lord Ganesha on a mouse, Kartika on Peacock, Laxmi on Owl and Durga on Lion/ Tiger *etc.* are symbols of endemic conservation practices. Similarly, promoting the use of flowers, leaves, grass and fruits of local origin during festivities are again designed to hallmark environmental conservation as a hologram of celebration.

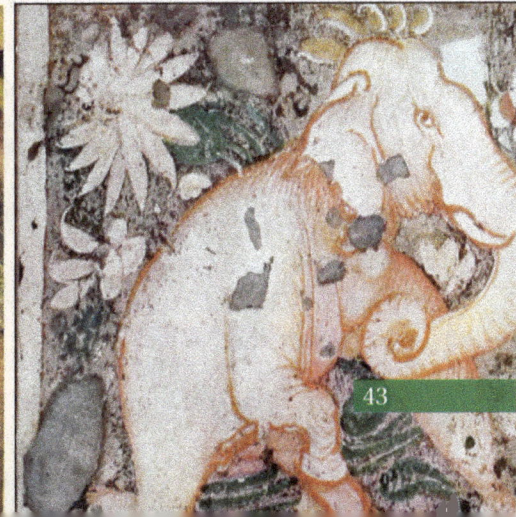

Environment as Capsule

24. Environment: Regional Wisdom and Ways

Irrigation

Along India's border with Bangladesh the people of Meghalaya use a 200-year-old irrigation system that directs water to their pepper plantations. The system is a low-tech answer to the wasteful industrial farms that use inefficient sprinklers and furrows. Using a set of bamboo pipes of multiple sizes the farmers direct water from natural streams downhill directly to the plants. The pipes are connected in a zigzag shape so that as the water runs through them it delivers a small but constant amount of water to each plant.

Terraced or Stepped Farming

Seen all over the world, from the Himalayas to South America, this agricultural practice has allowed cultures to harvest crops in places where farming might ordinarily seem impossible, on the sides of steep hills and mountains. It is called terracing and is a technique that has been used since Roman times. Farmers build a series of steps on the side of slopes using stonewalls or sod, and plant on the flat of each step. Aside from allowing crops to grow in difficult terrain, terracing is extremely water efficient. Each terrace step acts like its own reservoir. Without it, rainwater would run down the side of the hill carrying precious nutrient rich soil with it. More than being water efficient, this practice is time tested to halt the washing of topsoil that takes millions of years to be formed. One of the main problems of rivers in monsoon is the mud level that flows out of the unplanned conservation on hills and elevated landscapes. Nala plugging, terracing, stepped cultivation practices are our ancient solutions to forgotten current practices.

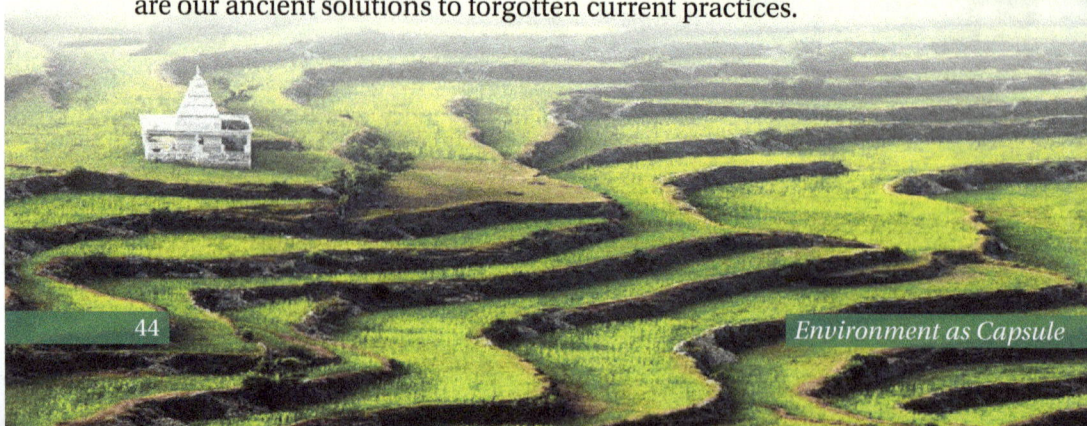

25. Three Sisters- Beans, Corn and Squash

The Native Americans use an age-old method. They grow three different crops beside each other to benefit the whole group—beans, corn and squash. They call the crops the 'three sisters'. The corn stalk provides a pole for which the bean vine to climb. The squash plant, whose wide leaves hug close to the ground, prevents weeds from growing, and acts as a mulch to retain moisture. The beans, which create nitrogen, provide nutrients to both plants. The plants even act in unison nutritionally–corn and beans provide the complete proteins we usually get from meat.

In contrast to the 'three sisters,' modern farms grow vast fields of identical crops called monocultures. Monocultures of corn for example erode topsoil and require large amounts of water and are more susceptible to disease. Polycultures, like the three sisters, behave more like the natural environment. In Western Ghats in Dang District of Gujarat, the tribal practice this polyculture by sowing Nagli millet along with oil seed crops (Kharsani) which helps to remove weeds, helps fertility of soil, attract honeybees during yellow flowering and this helps pollination process to give more millet yield and this sustainable practice saves soil, water and crops.

Constant use of pesticides and insecticides also furthers soil and air pollution and erosion. Repeated spraying of insecticides has led to disappearance of many species of birds, butterflies and animals apart from damaging the crop itself. It is also a frontrunner as the cause of increasing cancer cases in human beings.

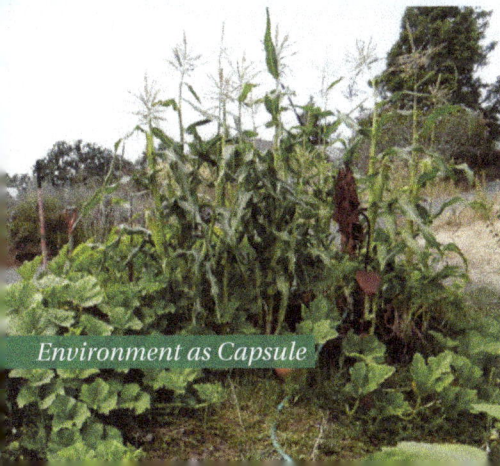

Three Sisters Garden: corn, beans & squash.

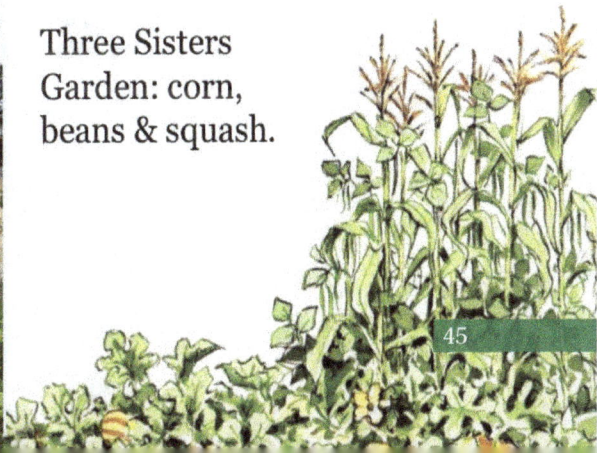

26. Major Challenges and Opinion

All of us should be proficient in Environmental education or environmental literacy.

The basic principles of ecology and nitty-gritties of environment can certainly help build a nous of earth-citizenship and a sense of duty to care for the Earth, its assets and to manage them in a sustainable way so that our children and grandchildren come into a harmless and unsoiled planet to live on.

- **Human values:** Manuals and resource materials about environmental education can play an imperative role in constructing optimistic attitudes about environment. The basic human value 'man in nature' rather than 'nature for man' needs infusion through the same.

- Mind is a mirror of a person- his inner feelings and attitude gives him tensions that cause bad health. Anger is Danger if D is the added to the result so we need to follow ancient "Sanatan" principle of living for others and getting delight in existence.

- **Social values:** We need to weave love, empathy, tolerance, patience, and fairness; that are the basic teachings of most of our religions, into environmental education. We also need to understand that, these values need nurturing so that all forms of life and the biodiversity on this earth are safe.

- **Cultural and religious values:** These are the ideals, preserved in Vedas like "Dehi Me Dadami Te" *i.e.* "you give me and I give you" (Yajurveda) that accentuate that man should not exploit nature without nurturing her. Our traditions and rituals, in many ways teach us that our actions must preserve, nurture and respect every aspect of nature because they are sacred. Be it rivers, earth, mountains or forests.

Environment as Capsule

- **Ethical values:** Environmental education should embrace the ethical belief of "earth- centric rather than human-centric" world-view. The educational system ou ght to endorse the earthcitizenship philosophy. Instead of considering human being as supreme, we must think of wellbeing of the earth. If the planet is not preserved and cherished, there will be no human being to enjoy his or his superiority.

- **Global values:** We need to understand and endorse the notion that, human civilization is merely a part of the planet as a whole. That nature and numerous natural phenomenon over the earth are interrelated and entwined with special ties of harmony and synchronization. If we interrupt this coherence anywhere, there will be an ecological disparity and subsequent catastrophe.

- **Spiritual values:** Ideologies like self-restraint, discipline, serenity, lessening of wants, independence from ravenousness and severity are some of the finest features convolutedly plaited into the traditional and religious fabric of our country. All these principles promote conservation and transmute our entrepreneurial approach.

The above-mentioned human morals, sociocultural, ethical, spiritual and global values incorporated into environmental education can go a long way in attaining the objectives of sustainable development and environmental conservation. Value based environmental education can bring in an overall transformation of our mind set, attitude and lifestyle.

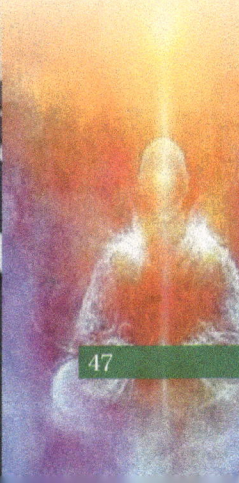

27. Practices to Teach Environmental Education

We need to instil the importance of environmental education through formal and informal methods to all sections of the society. Everyone needs to understand this, because 'environment belongs to all of us' and 'every individual and his actions matter' while conserving and protecting our environment. Several phases and techniques can be valuable for raising environmental cognizance in different sections of the society. Few of these are:

- **Formal Education:** Students should receive environmental education from the beginning. It is a welcome step that now all over the country we are introducing environmental studies as a subject at all stages including school and college level, following the directives of the Supreme Court.

- **Mass Media for Masses:** Media plays the quintessential role of educating the masses on environmental issues by printing articles, by organizing environmental rallies, by sponsoring plantation campaigns, via street plays, by talking about real eco-disasters and by sharing success stories of conservation efforts.

- **Government and peoples' representatives:** This elite section of our society that plays the most crucial role in deciding our future, it is therefore extremely important that they receive the required orientation and training. We need to make sure that they are well prepared and ready by organizing workshops and specially designed training modules. Publication of environment-related resource material in the form of pamphlets or booklets published by Ministry of Environment & Forests can also help in keeping this section abreast of the latest developments in the field.

28. Panchatatva and Humanity

Hinduism has several indications to reverence of the divine in nature, in Vedas, Upanishads, Puranas, Sutras, and many other sacred texts. Millions of Hindus recite Sanskrit mantras daily to venerate rivers, mountains, trees, animals, and the earth. Although the Chipko (tree hugging) Movement is the utmostac knowledged example of Hindu conservational guidance, there are multiple examples of accomplishments for welfare ofthe environment that are centuries old.

Hinduism is a remarkably diverse religious and cultural phenomenon, with many local and regional manifestations. Inside this cosmos of beliefs, a number of essential themes emerge. The diverse theologies of Hinduism suggest that:

- The earth can be seen as a manifestation of the goddess, and must be treated with respect.
- The five elements - space, air, fire, water, and earth - are the foundation of an interconnected web of life.
- Dharma - often translated as "duty" - can be reinterpreted to include our responsibility to care for the earth.
- Simple living is a model for the development of sustainable economies.
- Our treatment of nature directly affects our karma.

Gandhi exemplified many of these teachings, and his example continues to inspire contemporary social, religious, and environmental leaders in their efforts to protect the planet.

29. The Following are 10 Most Significant Hindu Teachings that are Prelude to Environment

Pancha Mahabhutas

The five great elements create a web of life that is shown forth in the structure and interconnectedness of the cosmos and the human body.

- Hinduism instils that the five great elements (space, air, fire, water, and earth) that constitute the environment are all derived from prakriti, the primal energy. Each of these elements has its own life and form; together the elements are interrelated and symbiotic.

- Hinduism acknowledges that human body is composed of and related to these five elements, and connects each of the elements to one of the five senses. Human nose is related to earth, tongue to water, eyes to fire, skin to air, and ears to space. This bond between our senses and the elements is the foundation of our human relationship with the natural world. For Hinduism, nature and the environment are not outside us, not alien or hostile to us. They are an inseparable part of our existence, and they constitute our very bodies.

Ishavasyam

- Divinity is omnipresent and takes infinite forms. Hindu texts such as the Bhagavad Gita (7.19, 13.13) and the Bhagavad Purana (2.2.41, 2.2.45), contain many references to the omnipresence of the Supreme divinity – including its presence throughout and with in nature. Hindus worship and accept the presence of God in nature. For example, many Hindus

- It should be in continuation with last para i.e. For example, many Hindus think of India's mighty rivers think of India's mighty rivers – such as the Ganges - as goddesses. In the Mahabharata, it is noted that the universe and every object in it has been created as an abode of the Supreme God meant for the benefit of all, implying that individual species should enjoy their role within a larger system, in relationship with other species.

Protecting the Environment is Part of Dharma

- Dharma, one of the most important Hindu concepts, has been translated into English as duty, virtue, cosmic order, and religion. In Hinduism, protecting the environment is an important expression of dharma. In past centuries, Indian communities – like other traditional communities – did not have an understanding of "the environment" as separate from the other spheres of activity in their lives. A number of rural Hindu communities such as the Bishnois, Bhils, and Swadhyaya have maintained strong communal practices to protect local ecosystems such as forests and water sources. These communities carry out these conservation-oriented practices not as "environmental" acts but rather as expressions of dharma. When Bishnois are protecting animals and trees, when Swadhyayis are building Vrikshamandiras (tree temples) and Nirmal Nirs (water harvesting sites), and when Bhils are practicing their rituals in sacred groves, they are simply expressing their reverence for creation according to Hindu teachings, not "restoring the environment." These traditional Indian groups do not see religion, ecology, and ethics as separate arenas of life. Instead, they understand it to be part of their dharma to treat creation with respect.

Our environmental Actions Affect our Karma

- Karma - a central Hindu teaching - holds that each of our actions creates consequences – good and bad – which constitute our karma and determine our future fate, including the place we will assume when we are reincarnated in our next life. Moral behaviour creates good karma, and our behaviour towards the environment has karmic consequences. Because we have free choice, even though we may have harmed the environment in the past, we can choose to protect the environment in the future, replacing environmentally destructive karmic patterns with good ones.

KARMA

The earth – Devi – is a goddess and our mother and deserves our devotion and protection

- Many Hindu rituals recognize that human beings benefit from the earth, and offer gratitude and protection in response. Many Hindus touch the floor before getting out of bed every morning and ask Devi to forgive them for trampling on her body. Millions of Hindus create kolams daily –artwork consisting of bits of rice or other food placed at their doorways in the morning. These kolams express Hindu's desire to offer sustenance to the earth, just as the earth sustains themselves. The Chipko movement – made famous by Chipko women's commitment to"hugging" trees in their community to protect them from clear-cutting by outside interests, represents a similar devotion to the earth.

Hinduism's tantric and yogic traditions affirm the sacredness of material reality and contain teachings and practices to unite people with divine energy

- Hinduism's Tantric tradition teaches that the entire universe is the manifestation of divine energy. Yoga – derived from the Sanskrit word meaning "to yoke" or "to unite" - refers to a series of mental and physical practices designed to connect the individual with this divine energy. Both these traditions affirm that all phenomena, objects, and individuals are expressions of the divine. And because these traditions both envision the earth as a Goddess, contemporary Hindu teachers have used these teachings to demonstrate the wrongness of the exploitation of the environment, women, and indigenous peoples.

Belief in reincarnation supports a sense of interconnectedness of all creation

- Hindus believe in the cycle of rebirth, wherein every being travels through millions of cycles of birth and rebirth in different forms, depending on their karma from previous lives. So, a person may be reincarnated as a person, animal, bird, or another part of the wider community of life. Because of this, and because all people are understood to pass through many lives on their pathway to ultimate liberation, reincarnation creates a sense of solidarity between people and all living things.

Through belief in reincarnation, Hinduism teaches that all species and all parts of the earth are part of an extended network of relationships connected over the millennia, with each part of this network deserving respect and reverence.

Ahimsa (Non violence) - Non-violence is the greatest Dharma. Ahimsa to the earth improves one's karma

- For observant Hindus, hurting or harming another being damages one's karma and obstructs advancement toward Moksha - liberation. To prevent the further accrual of bad karma, Hindus are instructed to avoid activities associated with violence and to follow a vegetarian diet. Based on this doctrine of Ahimsa, many observant Hindus oppose the institutionalized breeding and killing of animals, birds, and fish for human consumption.

Sanyasa (Asceticism) represents a path to liberation and is good for the earth

Hinduism teaches that asceticism – restraint in consumption and simplicity in living – represents a pathway towards moksha (liberation) which treats the earth with respect. A well-known Hindu teaching -Tain tyakten bhunjitha – has been translated, "Take what you need for your sustenance without a sense of entitlement or ownership."

One of the most prominent Hindu environmental leaders - Sunderlal Bahuguna – inspired many Hindus by his ascetic lifestyle. His repeated fasts and strenuous foot marches, undertaken to support and spread the message of the Chipko, distinguished him as a notable ascetic in our own time. In his capacity for suffering and his spirit of self-sacrifice, Hindus saw a living example of the renunciation of worldly ambition exhorted by Hindu scriptures.

Gandhi is a role model for simple living:

Gandhi's entire life can be seen as an ecological treatise. Here is one life where every minute act, emotion, or thought functioned much like an ecosystem: his small meals of nuts and fruits, his morning ablutions and everyday bodily practices, his periodic observances of silence, his morning walks, his cultivation of the small as much as of the big, his spinning wheel, his abhorrence of waste, his resorting to basic Hindu and Jain values of truth, nonviolence, celibacy, and fasting. The moralists, nonviolent activists, feminists, journalists, social reformers, trade union leaders, peasants, prohibitionists, nature-curelovers, renouncers, and environmentalists all take their inspirations from Gandhi's life and writings.

30. Vasudhaiva Kutumbakam, Global Village

Once we understand that everyone and everything we see is an expression and emanation of the Divine, we naturally embrace the globe as a village, Vasudhaiva Kutumbakam. Mother Earth "supports us with Her abundant endowments and riches; it is She who nourishes us; it is She who provides us with a sustainable environment; and it is She who, when angered by the misdeeds of Her children, punishes them with disasters." To Her, all are important.

31. Sarva Bhuta Hita, Welfare of All Beings

Once we understand Mother Earth's protection of life, we can understand how humans should act toward one another and all other forms of life. Thus, we arrive at Sarva Bhuta Hita, "enhancing the common good of all beings." When we know that all is sacred, all is God, and we are all children of Mother Earth, our behaviour and even our desires change accordingly. We want to enhance the common good and balance our individual needs with those of the extended family of life. It becomes natural to follow dharma. But even then, it is not necessarily easy to determine the best course of action for supporting the common good in specific situations.

32. Karma

An understanding of karma ties together these three grand concepts, informing us that our current condition is the combined product of our past actions (in this life and previous incarnations) and actions that we take today. Which means we are continually creating and contributing to our future, in the months, years, decades and even lifetimes to come. Evidently, our activities also impact and guide our family and community, today and into the future.

One of the most persistent environmental issues of our time is climate change. What did we do so wrong to cause greenhouse gas concentrations upsurge, so much that today it is the number one reason for countless climatic changes? Our parents, grandparents and great-grandparents (to a lesser degree) all burnt fossil fuels, cut down forests and increased cattle and poultry breeding to eat and sell meat, eggs and milk and milk products, without thinking about the consequences of their actions on environment and how would these actions shape the future for generations to come, for centuries. We only realized at the horrors we are about to face in the 20th century. Our ancestors, who were more religious, who believed in every word written in our ancient texts and believed it to be talisman of being; they created these frightening conditions that we face today and the future generations will have to bear the brunt worse than us. Environmental karma is the creation of human beings.

Still, we carry on unchanged and ignorant even today, despite the destructive effects, which have become apparent and continue to do so. It may be extremely hard to stop our harmful practices, not to mention reversing the damage we have already caused; however, we need to understand that this is not a danger that is somewhere in the future. This threat is staring at us right in our eyes, what corrective actions we take to deal with this, is what will decide a stable sustainable life or a life which is erratic. Who will suffer the worst of the environmental problems we have created? Certainly not our parents. Even those of us who are adults today may not bear the brunt of them. It will be our children, grandchildren and greatgrandchildren who suffer the consequences.

33. The Five Elements

In the Hindu conception of the cosmos and the environment, the five great elements (Pancha Mahabhutas) are central: space (Akasha), air (Vayu), fire (Agni), water (Apas) and earth (Prithivi). All emanate from prakriti (cosmic matter). Though each element has its own form and characteristics, all are interconnected and interdependent.

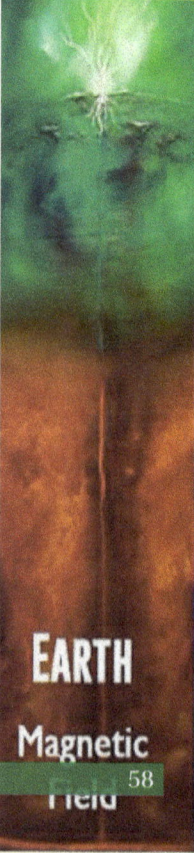

EARTH
Magnetic Field

WATER
Gravitation

FIRE
Solar Energy

AIR
Wind Energy

SPACE
Cosmic Radiation

Environment as Capsule

Akasha

Space, is the most subtle of the five elements, and there is no place where it is not. Akasha is not nothingness, like the popular conceptions of outer space; on the contrary, sky is absolute fullness. K. L. Seshagiri Rao explains, "Akasha represents openness, brightness, expansiveness and the fullness of blooming capacity."

Akasha is being thrown out of balance through excessive and constant noise, as is found in modern cities and towns with never-ending motor vehicle traffic, the hum of air conditioners and computer servers, blaring music, omnipresent advertisements appealing to the lower aspects of our being, televisions and video screens shoe horned into every available space. Needless to say, repose, reflection and sattvic (simple and minimalistic) living becomes difficult in such circumstances.

The Five Elements

In the Hindu conception of the cosmos and the environment, the five great elements (Pancha Mahabhutas) are central: space (Akasha), air (Vayu), fire (Agni), water (Apas) and earth (Prithivi). All emanate from prakriti (cosmic matter). Though each element has its own form and characteristics, all are interconnected and interdependent.

Vayu

Air, manifest on earth as the atmosphere, the shielding blanket of gases that environs the planet, regulating temperature and averting disproportionate solar radiation from reaching Earth's surface. Air unites and touches everything, from animals breathing in and out, to plants exchanging oxygen for carbon dioxide. Weather, which so defines daily life, is a function of air in collaboration with the other elements.

Unbalanced air is seen in pollution, particularly over cities. This is caused by industrial activity, power generation from fossil fuels, and the exhaust of internal combustion engines. Brown smog is the most visible form of air pollution, but an excess of greenhouse gases also affronts air, causing climate change, raising temperatures and slandering oceans. Any disturbance to air can also interfere with another element-Agni, as in its solar form Agni is the energy which heats the air, powers the water cycle, and helps enable life to grow. Each of those processes is affected by pollution. Planting wide leaves trees can act as dust catchers other than giving oxygen cycle to homes.

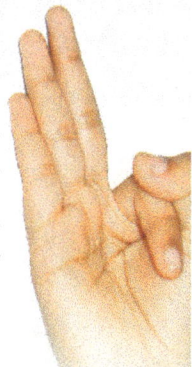

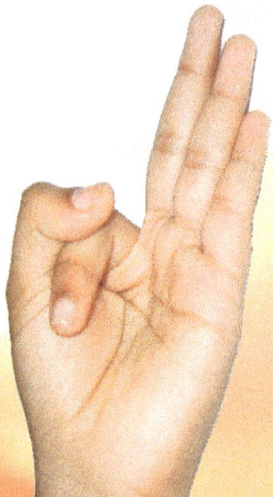

Agni

Fire, has been worshiped since ancient times. Fire purifies, fire destroys, fire inspires. From the Sun, to lightning, to fire in its mundane and sacred forms, Agni brings warmth and visibility to the world. The Vedas sing its praises: "I magnify the Lord (Agni), the divine, the priest, minister of the sacrifice, the offeror, and supreme giver of treasure. To you, dispeller of the night, we come with daily prayer, offering to you our reverence" (Rig Veda 1.1.1&7).

Agni

Water

Apas, is the source and retainer of life. Its immense sacredness is rivalled only by its practical value to human agriculture, health, enjoyment and the development of civilization. In the form of Earth's rivers, water is so vast in its life-giving and life sustaining properties that it is worshiped as the mother of life, as Mother Ganga.

The element water has been drastically disturbed by human activity. We have dumped human sewage and industrial effluent into rivers and oceans; farmers have irrigated so aggressively that water tables have become lowered, and have allowed runoff of chemical fertilizers and pesticides into streams; we have built massive dams and diverted entire rivers at the expense of wildlife and the people who live in the watersheds; and we have disposed of nonbiodegradable trash and plastics in rivers, ponds and wetlands. The same activities that unbalance vayu and agni, causing climate change, also affect water, causing ocean acidification, coral bleaching and glacial melting.

Earth

The impermeable of the five elements, is the ground upon which life takes shape and form. It is the body of the Divine, a living organism, metaphysically, metaphorically and biologically. Hindus have cognized this for millennia, knowing that all creatures are intimately connected to the Earth. Without its gifts, we are nothing. In the latter half of the 20th century, modern science has caught up with this truth, conceiving of Earth as a self-regulating organism called Gaia, a name for which Dharani, Bhudevi or Bhumi-Hindu names for the Goddess-could easily be substituted.

Earth unbalanced is the most obvious. Enormous expanses of forest are cleared for timber and agriculture-often for industrial farming or cattle ranching. Mining for coal, bauxite, gold and hundreds of other minerals blemish mountainsides and eradicate mountaintops. Human demand for resources are awfully unbalanced and unequal between rich and poor nations and resulting in habitat loss and species extinction at a level unprecedented in the modern history of the planet. Human population, which is rapidly increasing only endangers other species furthermore.

To sustainably stream out our existing resources, consumption would require one and one half planets. By 2030, that is estimated to rise to two planets. We can hope that a new balance will sooner or later be reached; but ahuge majority of climate scientists say that in that new balance, the Earth will be less fertile than now. Earth has endured deep-seated changes many times. Ice ages come and go in anticipated cycles. Species are destroyed and new life emerges. Temperatures waver, oceans rise and fall. This is not the first challenge to life Earth has faced, but it is the first in which we, the human race, are playing a major role.

Though self-regulating and dynamically interacting, the five great elements of existence can be pushed out of balance by human action, creating conditions inhospitable to life in general. What are the imbalances confronting the Earth?

34. It is probable to Bring Forth a New Era

There is the doctrine of the four yugas. We are living in the Kali Yuga, which started around 3102 BCE. This is the worst of all the four ages, and things are going to go downhill. However, there is a very important point which is often overlooked. The same texts that talk about these ages, like the Manusmriti, also say that the king is the maker of the age. That is, the sequence of the ages can be reversed by a political initiator, to put it in modern idiom. This is found in the Mahabharata and in many scriptures from ancient India in which the king said that he established the golden age in this age of the Kali Yuga. So here we have a very clear provision of intervention to prevent environmental degradation, especially by the state.

35. God is Everywhere

Hindu philosophy is based on the truth that there is one Supreme Power that is the sustaining force of the entire creation. Personal transformation starts with realization of this Supreme Power within one's own self. The aspirant will then be able to experience that power all around him. Thus he understands that this power is universal, non dual, indivisible and eternal. He sees unity in diversity. He will not see his fellow human beings as different from him and so does not fear. Such a person is full of compassion and unlimited love. He will work towards peace and prosperity of not only mankind but all of nature. This is accomplished only through faith and surrender to that Supreme Power and under the able guidance of the spiritual teacher who is an embodiment of that power. Healing the Earth is possible by exchanging ideas and restoring spiritual values. Peace is much needed in today's world. Unless each individual changes his behaviour and thinking for his own progress and for the world at large, peace cannot be established.

36. Stop All Killing & Exploitation

Reverence is the root of Hinduism, reverence for all life. People worship trees, they worship mountains, and they worship the universe, the spiritual world. Now it is all gone. Hinduism can save the world from global annihilation. Hinduism has the potency and the power. But today people don't pay attention to our great ones. No sentient creature must be killed. If I kill an animal for food, I will not hesitate to kill a fellow human being whom I regard as an enemy. Human race will learn one day that there is no other way to peace than vegetarianism. Life is a gift of God. Among all the creatures, man is the only one who has been given the power to meddle with the ecological balance. Therefore, he has great responsibility to see that all types of life are preserved. All life should be regarded as sacred, for there is but one life that flows into all. This one life sleeps in the mineral and the stone. This one life stirs in the vegetable and the plant. This one life dreams in the bird and the animal. This one life is awake in man.

37. What will it Take to Avert Dire Climate Change?

The average per-capita carbon footprint for people living in the United States is approximately 18 tons per year, though it has fallen in the past few years due to recession. The average in Europe is about half that, with China coming in at about 6 tons. In India the number drops to roughly 2 tons. If the goal is to keep temperature rise below 2 degree celsius- the threshold above which many dangerous climatic changes are said to become unavoidable-and if we accept that every human has the right to similar levels of development, then 2 tons is roughly what each human being needs to produce. In this one statistic the enormity of combating climate change becomes clear.

38. Hindu Virtues versus Consumerism

Hinduism's numerous classic restraints and practices, the yamas and niyamas, offer heaps of applied direction for those wanting to minimize their influence on the environment. If we are to observe non-stealing, we cannot use natural resources at unsustainable rates; when we do so, we endanger the quality of life of future generations. This is effectively a form of stealing. If every human on the planet consumed resources at the level of the United States, we would need 4.5 earth to supply everyone's needs. That decreases slightly if everyone lived like the average European or Japanese citizen, but still not to ecologically sustainable levels. To reach that point, consumption like that of the average Thai citizen, or less, is required. If more people practiced gratification, we would greatly reduce the impact of consumerism on the planet-from rising greenhouse gas emissions, to chemical pollution, e-waste from countless short-lived electronic products, and the myriad consumer goods that get used and thrown away each year in the wealthy and growing nations of the world. Contentment involves living in constant gratitude for your health, your friends and those belongings which you do own, not complaining about what you don't possess, as well as viewing every moment in life as an opportunity for spiritual growth and development. All of this has great positive environmental impact. Being contented in the moment, living in the eternal now, insulates you from consumerism, allowing you to embrace a simple life, caring for what you have and living within your means.

Having Concern for Others

The approach that we should take the maximum wealth available to us from nature, be it oil or metals, and that we should maximize our power with nuclear weapons-these contribute to our global problems. Our Hindu dharma has given us certain important values to implement in our day-to-day lives, including being satisfied with whatever we have. Learn to share, learn to give first, and then enjoy. This attitude will bring about harmony in society. Hinduism is a tradition which has always cared for the growth and religious sensitivities of each and every individual-not only cared for but helped them equally to grow individually. Today the absence of this attitude has created agitation and given rise to crime and imbalance in society. The attitude that "I shall grow at the cost of others" is considered improper in the Hindu religion. It is a great sin against ahimsa, the principle of nonviolence, to be insensitive to the rights and demands of others and to afflict pain or hurt on them-not only physically, but by hurting their religious sentiments, their belief systems.

39. Personally

While some environmental issues seem beyond the control of the individual, there is still much a person or a family can do. Many lists of things you can do to green your life focus on myriad small steps, such as recycling. But to get an overview, there are three important areas on which to concentrate: what you eat, how you use energy and how you get around.

Your Diet: The environmental benefits of being vegetarian, particularly when you also eat organically grown produce, are numerous. A vegetarian diet reduces your personal carbon emissions by over one ton per year, compared to someone who eats meat, while a vegan diet reduces it even further.

Your Power: If your electricity supplier offers an option to use renewable energy, choosing this is a great way to lower your home's environmental impact. Whether you have such an option or not, there are many ways to reduce your power consumption and thus be more Earth friendly: go solar, improve insulation, install timers and motion sensors, air-dry your clothes, use rechargeable batteries, turn off lights, use on-demand gas water heaters and LED lights, hand tools rather than power tools, *etc.*

Your Transportation: Choose the least damaging way of getting from point A to point B, the one with the lowest carbon footprint. Aviation, for example, is hugely energy intensive. Just one long flight a year, say New York to Los Angeles or London, nearly equals in carbon emissions the entire yearly emissions of the average Indian citizen. A train or bus creates a small fraction of the pollution. Choose a car with high fuel efficiency and share the ride with others. For short daily journeys, walking and bicycling are by far best for the environment- and for your health and finances.

40. Insights from the Vedas & Ayurveda

The upanishads teach us that everything is Brahman ("sarvamkhalvidam Brahman") or Satchidananda, Being- Consciousness-Bliss, differing by apparent names and forms only, not by essential nature. This does not mean that God created the world, but that God and the world are one as visible and invisible facets of the same ocean of mindfulness. All life is not merely interdependent but is one at its core with the Supreme Truth.

This Vedic honouring of the sacred nature of all life is called yajna, sometimes translated as "sacrifice," but which really refers to a sacred way of life and action that recognizes the divine presence in all things and strives to live in harmony with it. Our lives should be a ritual in which we strive to pursue a way of right action in harmony with the rhythms of nature and of the spirit through which nature works.

The practice of yoga arose as the inner sacrifice, or antaryaga, the offering of speech, breath and mind into the divine flame of awareness or Agni within our hearts. The yoga asana itself is meant to establish a sacred connection with the Earth. Yoga itself should be a sacred art of communing with all of life.

Environment as Capsule

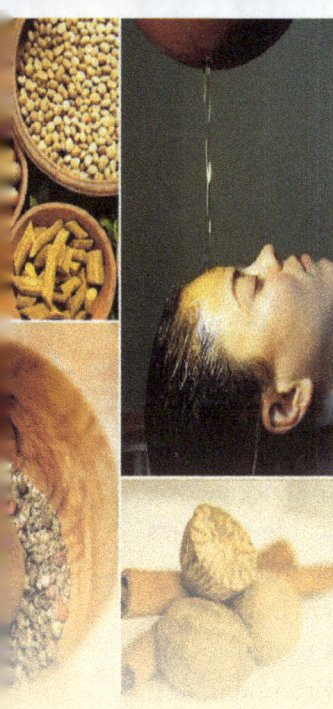

Ayurveda

Cautions of prevalent diseases, both physical and psychological in nature, which can ascend through destruction to our environment. The Charak Samhita (III.6.23) discusses the diseasecausing effects of polluted and disturbed air, water, land and seasons as a root of obliteration of all countries. Twenty-eight factors of damage to air, water and land are listed, of which we can find all taking place in the world today. Further harmful factors to the outer world, they include obstinate and self-regarding behaviour on the part of human beings, their fall from ethical behaviour and indifference for spiritual practices, particularly unrighteous conduct by the rulers of a country and, above all, violence and war.

Charak states that when natural time cycles like the seasons are disturbed, the situation will become precarious. Yet he also states that such collective problems and diseases can be and countered by health practices like Panchakarma, by leading a simple minimalistic life and the practices of yoga and meditation. Clearly our disruption of the environment has consequences both of a material and spiritual nature, though these may take some more decades to fully manifest, as nature works on a slower time cycle than human beings. We must reconnect ourselves with universal peace and once more come to honour the Earth and nature in order to solve this dire situation.

41. Natural Degradation of Stuff We Use

From cars to food wrap and from planes to pens, you can make any thing and everything from plastics—unquestionably the world's most versatile materials. But there's a snag. Plastics are synthetic (artificially created) chemicals that don't belong in our world and don't mix well with nature. Discarded plastics are a big cause of pollution, cluttering rivers, seas, and beaches, killing fish, choking birds, and making our environment a much less attractive place. There are hundreds of man-made products which are easily degradable in environment and cause no harm to the nature, sustainable way. In recent times, a new term called 'biodegradable' on some products that we buy, such as washing powders and shampoo, have evolved, but what does it actually mean?

The 'degradable' part of the word simply means that the product is able to break down into smaller compounds and eventually into very simple compounds such as carbon dioxide, water and oxygen.

The 'bio' part of the word means that the process is helped along with biological organisms, such as fungi and bacteria, which digest the material.

So a 'biodegradable' object is one that will break down quickly and safely into harmless compounds by using the action of micro organisms.

Composting is another name for bio-degrading. When a leaf falls off a tree onto the ground, it will rot away until nothing visible is left – this is bio-degradation at work. Fungi and bacteria break up the leaf into smaller and smaller parts and eventually into carbon dioxide, oxygen and other important elements that have helped make up the leaf, such as phosphate and nitrogen. These simple elements are harmless in the environment and can actually be beneficial, as it enables the tree to get back some nutrients. This is why people often collect and compost their biodegradable waste and later, when the material has broken down, put it on to the garden to act as fertilizer.

So what makes a product 'biodegradable'? Any material that comes from nature will be able to return to nature provided that it hasn't been changed too much by manmade processing. For example, we all know that an apple core is capable of bio-degrading even if we accidentally leave it in our school bag, but what about a T-shirt? A Tshirt is made from cotton, which is from a plant, but it has been processed by spinning the thread and weaving the cloth – but because it is still basically from a plant, it will break down, though not as quickly as it would had the fibre been left on the plant.

Any plant-based, animal-based or natural mineral-based product has the capability to bio degrade, but they will bio degrade at different rates depending on the original material and how much it has been processed.

Products made from man-made compounds formulated in a laboratory are in combinations that do not exist in nature and therefore microorganisms are not able to break them down. Therefore, we should be able to understand the entire issue of natural degradation and adapt products or articles that can be degraded in the nature within a certain time.

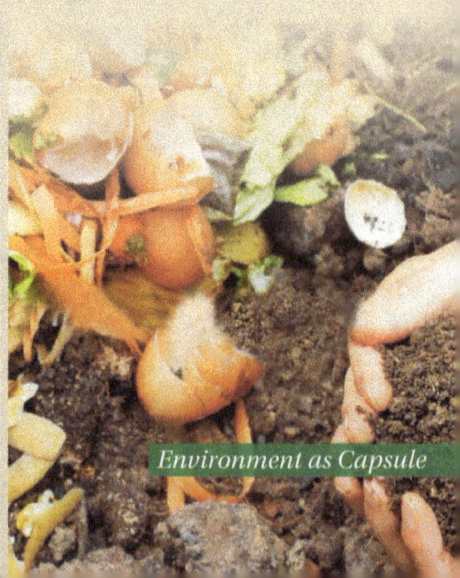

42. Bio Indicators - Best way to Learn about Environment

A bioindicator is a living organism that gives us an idea of the health of an ecosystem. Some organisms are very sensitive to pollution in their environment, so if pollutants are present, the organism may change its morphology, physiology or behaviour, or it could even die.

One example of a bioindicator is lichens. These plants, which live on surfaces such as trees or rocks or soil, are very sensitive to toxins in the air. This is because they obtain their nutrients mostly from the air. We can tell our forests have clean air by the amount and types of lichens on the trees. Different species of lichen have different levels of susceptibility to air pollution, so we can also get an idea of the level of pollution by observing which species are present.

Bioindicators can be plants, animals or microorganisms:

- If toxins are present, certain plants may not be able to grow in the area affected.

- Monitoring population numbers of animals may indicate damage to the ecosystem in which they live.

- Algae blooms are often used to indicate large increases of nitrates and phosphates in lakes and rivers.

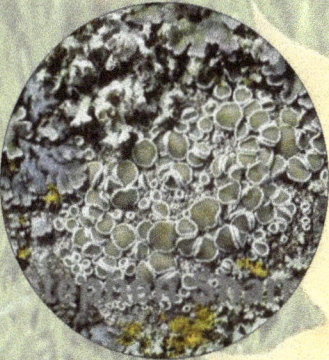

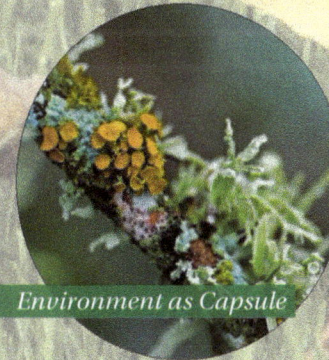

If pollution causes the reduction of an important food source, the animals dependent on it for food may also decrease. Animals may also change their behaviour or physiology if a toxin is present. For example: the levels of certain liver enzymes in fish increase if they are exposed to pollutants in the water changes in the functioning of the nervous systems of worms are used to measure levels of soil pollution the increase in the number of mutated frogs found in the USA is used as an indicator of toxins in their environment.

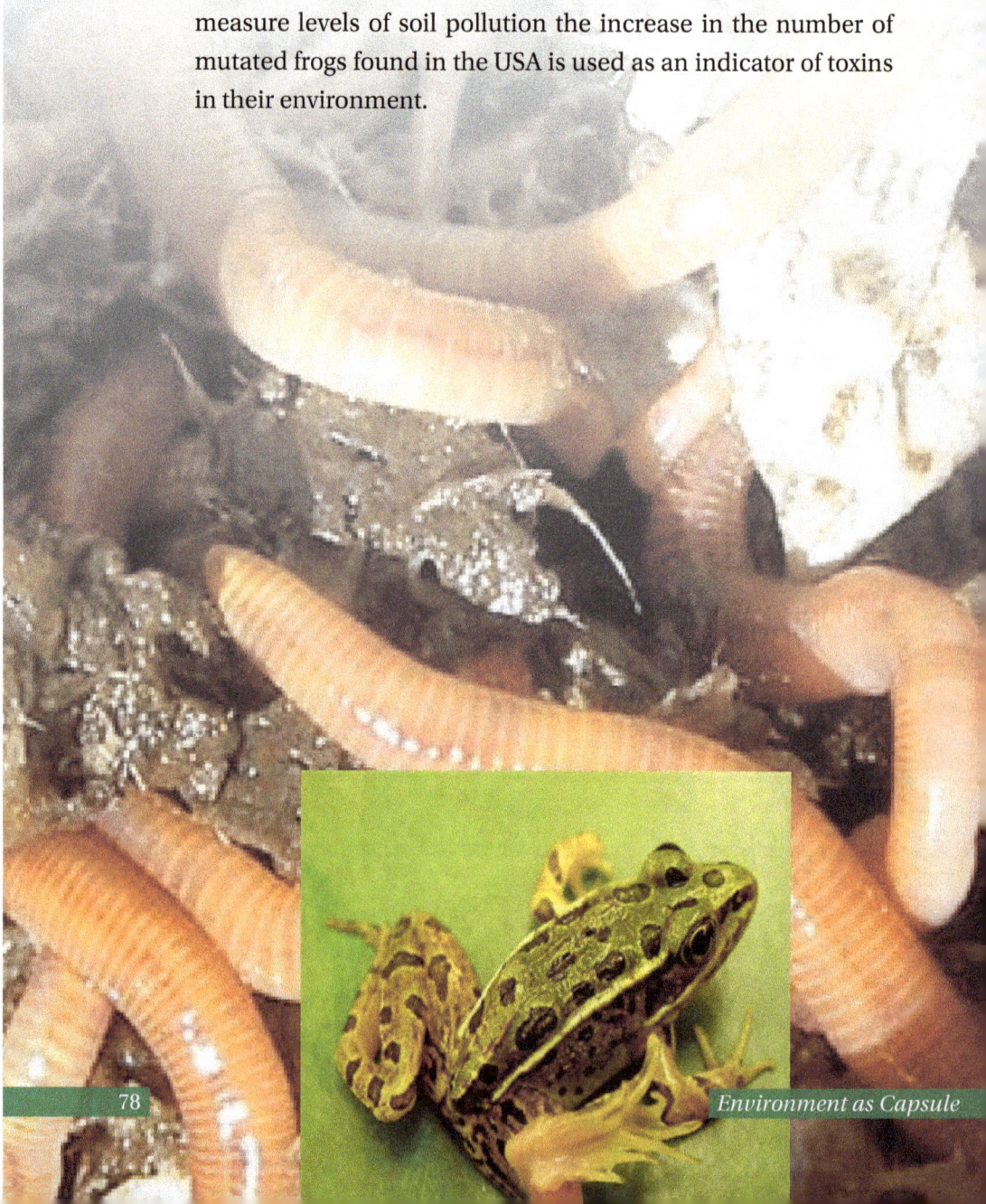

Nature's Signalling of Weather System

Long before the initiation of modern scientific methods for weather forecasting and climate prediction, farming continued successfully, with the exception of regular disasters. Farmers utilized traditional ways and indicators of rainfall forecasting/prediction. Compared to the dominant industrialized societies, in which activities in the last 200 years or so have caused most of the climate impacts currently observed, indigenous people living on their traditional lands bear little responsibility for current and future projected consequences of a changing climate. Despite this, they are likely to suffer the most from direct and indirect climate change. Following are the few age-old tips which helped forest dwellers to predict weather.

Common Frog: If the frogs croak in a water body in the afternoon until sunset, rain will be coming soon, even during winter and spring season.

Bird/Hen and cock: If local domestic chickens search for food even during the rain, it is commonly thought that the rain will last for the whole day. But if the birds stop searching for food when it is raining and take shelter (in the morning or afternoon), the rain is expected to cease soon and to be minimal.

Location, pattern of clouds (blackish colour): When the clouds are thick and black in colour, and are arranged perpendicular to the orbit of the sun in the morning, it is said that rain is approaching.

Cloud colour, time, direction and location
of appearance in the sky

If a reddish colour cloud is seen at sunset one western horizon, rain is predicted to come within two to four days. If there is thick cloud toward the south or north, the rain will be more on that side within the region. If the sky is full of reddish coloured clouds appearing after a long rainfall, it is a sign that the rain will not come again in that particular season.

Rainbow Colours: red dominating means more rains to come, if blue colour dominates and clear sky appears it means that rain has passed. Stratus cloud is a sign for cold days.

Hawks flying high means a clear sky. When they fly low, prepare for a blow.

Birds tend to stop flying and take refuge at the coast if a storm is coming. They'll also fly low to avoid the discomfort of the falling air pressure. When seagulls fly inland, expect a storm.

Birds tend to get very quiet before a big storm. If you've ever been walking in the woods before a storm, the natural world is eerily silent! Birds also sing if the weather is improving. Birds singing in the rain indicates fair weather approaching.

Dropping of fruits before maturity indicates very dry season or drought must be expected. Immature fruits drying on trees and/or dropping from the trees is an indication of drought. Moon crescent facing upwards indicates upholding water and when facing downwards is releasing rainfall in next three days.

Well-fed calves jumping around happily in the field and on their way home from grazing in the mountains and unwilling to graze the following morning indicates good rains on the way

Increased libido in goats and sheep with frequent mating is a sign for good rains.

Ants queuing up for space protected areas would signal the beginning of wet spell.

43. Major Initiatives by Government of Gujarat

Ganesh Idol Immersion

Ganpati festival has its share of ecological concerns. The rivers, which are already exposed to a lot of pollution, are now subjected to the vile material that Plaster of Paris is.

Why eco-friendly idols? Plaster of Paris is fine white powder made by calcifying gypsum, it is then mixed with water to make a thick paste that can be cast into moulds and made into any shape you want. It later hardens without shrinking or even cracking. Undoubtedly, the idols made with this material cost much less and the finishing are finer than the clay or mud idols. The idols made using PoP are painted using toxic chemicals, which when immersed in water, harm the ecosystem to dangerous levels. The synthetic chemicals used in the paint and the PoP spread into the water and pollute the river, increase the acidity of the water and kill a lot of aquatic life while the rest of the remnants get washed back onto the shore within hours. In short, you are basically releasing a container of chemicals inside the water, in the name of an idol. Alternatively, a clay, paper or mud idol will simply melt and blend into the waters.

If you think that getting a clay or mud idol custom made for you is heavier on your pocket that you would like to spend an easy alternative is to use paper mache or small terracotta (baked clay) idols. These are easily available in most craft stores across your city and will not be as expensive as the clay ones either. If that doesn't work for you, here's a smart suggestion. A lot of people don't make the idol customary for the festival. The celebrations are more about welcoming and hosting Lord Ganesha in your hearts and homes. It is possible to completely do away with the idol and place a beetle nut (Supari) on the pedestal and pray to it instead. Some people who want to feel the festival in its entirety also place their own Ganesha idol on the pedestal with the Supari and during the time of the visarjan, immerse only the Supari, replacing the idol back in their temple.

This shows that if there is a will, there definitely is a way! So next time during Ganpati festivals, don't make the rivers cry, let it be a fun fest for everyone!

In its resolution, the Gujarat Government directed that the idols should be made from natural materials as described in holy scripts. Idols should be made of traditional clay and not from use of baked clay, Plaster of Paris. PoP idols don't dissolve completely in water and cause cloggs and other problems in the river or other reservoirs where people immerse it during the festivals. To avoid environment degradation, the government has proposed clay idols and also a separate tank for immersing idol.

Banning plastic ropes for kite flying

Through a drastic implementation strategy in 2009 by use of Section 5 of EP Act, the production, sale, trading, storing and purchase of plastic ropes (of Chinese origin) were banned since all these would not only be non-degredable but by virtue of its wide reach across the sky could clog water ways, cause injuries, disturb the soil and agriculture operations. Keeping this factor in mind, a sound environment practice to use cotton ropes was reactivated.

Acceptability of CNG

Around 5 years back, there are more than 64870 CNG auto rickshaws in Ahmedabad; out of which 51894 were new rickshaws. Total 88854 vehicles in the city were on CNG, and 32784 were on LPG. In parallel, to augment supply, total 63 CNG stations were operational in Ahmedabad that time. The Ahmedabad Municipal Transport Service (AMTS) had put 533 CNG buses on road along with 50 low floor Euro -III Diesel buses; 650 more feeder buses added later. They had decided to scrap the old diesel buses (283) in phased manner by the end of the year in 2011. Today, the entire fleet of buses on Ahmedabad- Gandhinagar route of Gujarat State Road Transport Corporation (GSRTC) are CNG fuelled. The concept of CNG has been greatly accepted by the people of this city. One can observe the change now.

44. Environment and Public Policy

Granting of sand mines or stones in rivers could cause severe imbalances unless proper study of the river course or deposits of such type causing obstruction are not established.

Granting of permission to use water of reservoirs located in sanctuaries or forests could be a regressive policy measure in terms of Environment and Conservation.

Tourist resorts or hotels that get located far away from the water source based only on locations advantage could more counterproductive since not only water depletion would accrue at originating point out also avoidable energy consumption would follow suit.

Permitting chemical units in fragile water zones would be disastrous in terms of source pollution while locating these across coastal tracts to diffuse pollutants as per admissible dissolution would be more welcome.

Removing trees while expanding or broadening of roads should be the last option while preserving them as an avenue and making roads to trail such avenue formation would be more suitable and rewarding.

45. Do's & Don'ts

Let's Start from Home

Do's

1. Use mug instead of running tap while brushing teeth.
2. While watering plants, instead of running hose, use water cane.
3. Use a toilet flush which consumes less water.
4. Carry cloth, jute or paper bag to the market.
5. Use dustbin for garbage disposal.
6. Plant a garden. Even in urban settings, you can grow herbs and flowers in pots.
7. Wear extra-layer of clothes at home instead of turning up the heater.
8. While shaving, use mug instead running hose.
9. If you have a choice, pick paper bags over plastic bags at the grocer and elsewhere if you do take plastic bags, wash and reuse them.

Don'ts

1. While taking bath, don't use shower run for long.
2. Don't allow water overflow from the over head tank.
3. Don't buy loud crackers during Deepawali.
4. For washing floor, don't use running hose, use mop and bucket.
5. Don't junk things break rather think to fix them.
6. Avoid unnecessary use of lights and fans.
7. Don't louder the volume of your TV, radio and music system.
8. Never leave food residue in your plate uneaten.
9. Don't over packaged goods and foods. Containers and packaging make up about a quarter of the waste stream.

Prevention and Control of Vehicular Pollution

Do's

1. Do you really need to drive a car everywhere? Walk to work, or ride a bicycle.
2. Carpool. Two – or four – can ride as cheaply as one.
3. Get a valid pollution under control certificate from authorized testing centre.
4. Clean up your act. Keep automobiles fuel filters clean and save the fuel.
5. Clean the air filter and oil filter regularly.
6. Clean the carbon deposit from silencer.
7. Maintain recommended tyre pressure.

Don'ts

1. Don't use extensively your private vehicles, try to use public transportation whenever possible.
2. Avoid congested road and rush hours.
3. Don't idle away energy. Beyond one minute, it is more fuel-efficient to restart your car.
4. Don't forget to keep your vehicle tuned up. When a vehicle is running well, it uses nine per cent less fuel and thus emits fewer toxic and noxious fumes.
5. Don't try to replicate mechanical works and experiment with your car.
6. Don't forget to replace your old battery with new battery when it required.
7. Don't use clutch pedal as footrest.

Prevention and Control of Water Pollution

Do's

1. Reuse of water whenever possible, kitchen water can be used for watering the plants.
2. Plan your kitchen activity to avoid wastage of fuel and water.
3. Fix leaks promptly. A dripping joint can waste more than 76 liters of water a day.
4. Take showers instead of baths. Showers use less water, if you limit them to five minutes.
5. Install low-flow shower-heads.
6. Use sprinkler for irrigation.
7. Use scientific method of application fertilizers.
8. While shaving, use mug instead running hose.
9. Run your dishwasher, washing machine and dryer only when you have full loads. When possible, use an outdoor clothesline instead of a clothes dryer.

Don'ts

1. Don't keep on the tap running while having bath, brushing teeth or washing dishes. It wastes about 2 liters of water every minute.
2. Don't hose down your lawn or corridor to clean it. Sweep it off.
3. Don't wash the clothes and kitchen utensils in the water bodies.
4. Don't litter. When camping, keep the areas clean.
5. Avoid throwing flowers, sweets, puja materials into a river. It will degrade the quality of water. The river won't be happy with this.
6. Avoid throwing dead bodies in a river. This will ultimately landing in the mouth of dogs, vultures & other animals. Do you like it to happen with your bodies.
7. Never dump anything into the water bodies.
8. Avoid use of weedicides.

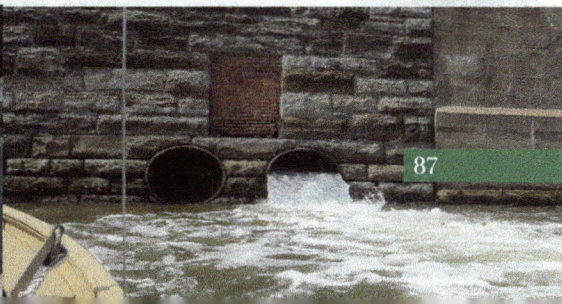

Prevention and Control of Noise Pollution

Do's

1. Always maintain your motor vehicle and its exhaust silencer in proper condition.
2. Ensure that your Diesel Generator Set is provided with acoustic enclosure which gives a reduction of a minimum 25 dBA (as per the provisions of the Govt. of India notification GSR 371(E), dated May 17, 2002.
3. Ask your copy of valid type approval certificate from the dealer while purchasing portable petrol/kerosene generator sets (as per Govt. of India notification viz. GSR 742(E), dated September 25, 2000, which prescribes noise standards for petrol/kerosene generator sets).
4. Keep the volume of the loudspeaker or sound amplification system low so as not to annoy your neighbours.
5. Ensure that the sound from your music system is played at volume which does not disturb your neighbour.
6. Play free-crackers only outdoors in large open areas and community level.
7. Purchase only those fire-crackers that comply the noise standards as provided by the Govt. of India regulation GSR 682(E), dated October 5, 1999.

Don'ts

1. Avoid using horns except at emergencies.
2. Avoid use of multi toned/air horns in your vehicle.
3. Do not install Diesel Generator Sets without prior approval of the competent authority, if required by local laws.
4. Avoid use of loudspeaker in the open.
5. Do not use loudspeaker or any sound amplification system between 10:00 P.M. and 6:00 A.M., except in closed premises.
6. Do not make your neighbour a captive listener to your music system.
7. Do not play fire-crackers between 10:00 P.M. to 6:00 A.M.

Energy Conservation

Do's

1. Clean the condenser coils on the back or bottom of your refrigerator once a year. Adjust refrigerator to a less-cold setting.

2. Make maximum use of natural light.

3. If you use a radiator heat, put a reflector sheet behind the radiator. It keeps the heat from being absorbed by the wall.

Don'ts

1. Don't keep on the lights if you will be out of the room for 15 minutes or more.

2. Don't use electricity during day time. Think about it – do you really need to turn on a light during the day.

3. Don't install Diesel Generator Sets without prior approval of the competent authority, if required by the local laws.

4. Don't prefer incandescent light and replace energy efficient, compact fluorescent light tubes even if it costs more.

46. Conclusion

Science, technology, and indigenous knowledge put together are the key to conserving ecosystem and leading a sustainable life. Ancient medicinal texts may still hold answers to treating the most dangerous diseases that have gained momentum because of climate change.

Equity of knowledge provides opportunity for local people to participate in management of local affairs with global implications. It also provides the opportunity for self-determination and self-preservation. By accepting and using indigenous knowledge, we shall uphold the code of parity of knowledge. Evenhandedness of knowledge between local and formal sciences will empower, secure and provide opportunities to locals. By incorporating formal and local knowledge to devise and implement resource management, the state will lessen the societal hurdles in participation and augment capability of indigenous people to seek solutions for problem solving. The affluence of knowledge amassed by traditional societies shall be a gift inherited by generations to come. The knowledge of conserving water, forests and other natural resources will help us make the natural resources last longer. We will be able to find elucidations to bigger issues like global warming and succeed in diffusing the potential of further damage to the ecosystem caused by depleting natural resources. Comprehensive and combined astuteness will help us design and contrive suitable programmes to manage agro and ethno forests. This will enable us to warrant ecological, economic, and social refuge for younger generations.

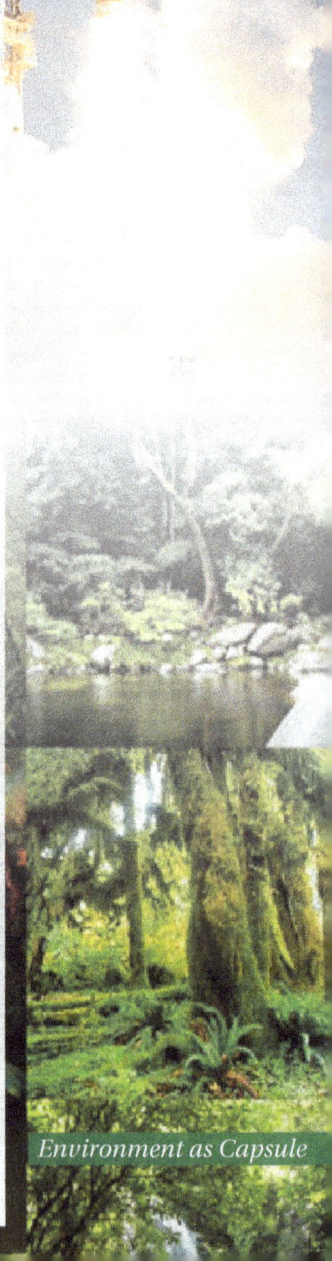

The 1992 Convention on Biological Diversity requires that every Contracting Party respect, preserve and maintain knowledge, innovations and practices of traditional and local communities and promote the wider application with the approval and involvement of the holder of such knowledge, innovations and practices and encourage the equitable sharing of the benefits.

Formal science assures augmented productivity and efficiency in managing natural resource by following the process of acquisition, transmission, integration, and field application of traditional knowledge on afforestation and planting trees in urban spaces.

Importance of distinguishing valuable knowledge from myth has become imperative. This will prove valuable from a different viewpoint too: the valuable knowledge is preserved. Identification of science behind traditions is a more constructive effort than entering into the 'indigenous vs. scientific' or 'traditional vs. western' arguments. Scientists can be as critical as they want; just as local people do not approach formal science without apprehensions. Persuasive promoters of both local and formal knowledge system have brought enough harm by protecting exclusive claims to prove their loyalty towards the system they believe in.

Nonetheless, formally trained scientists as well as researchers on traditional knowledge systems misconstrue the process of validation. The term 'validation' does not mean a "narrow reductionist perspective of disciplinary confines." It can, and should, appeal on complementarity and consilience across local and formal systems. Consequently, both formal and local methods, as well as local people and formally trained scientists, intend to contribute and add value to grasp the data, information and knowledge. When we start advocating and empowering such comprehensive and collective efforts, we shall mutually benefit. Unlimited access to recent developments and advancements in science and technology will help improvising existing norms and practices of conservation. Our local knowledge is rich because it is based on 'Experience' while formal modern stream relies on 'Evidence' and protocols through trials which may not hold good for the thousand and more remedies in traditional system.

Environment as Capsule

Local and formal knowledge have both benefitted by accepting each other and by recognizing the interactions between nature and society. There are abundant instances of local knowledge being expedient in corroborating scientific propositions and proposing new research guidelines. Similarly, formal scientific approaches have been extremely valuable in certifying traditional ethno-pharmacological knowledge by ascertaining the vital constituents (chemicals) in plants used in ancient medicines. One noteworthy input that established the ancient-modern concordance came with the isolation of the hypertensive alkaloid from the Sarpagandha plant (*Rouwolfia serpentina*), esteemed in Ayurveda for treating diseases like hypertension, insomnia, and insanity. There have been quite a lot of such contributions since then.

We need to do away with the vain theoretical wiles and debates on sovereignty of one belief over the other, we are at a juncture today, when going above and beyond this sociocultural boundary and reaching a mutually beneficial ground across ethos, faith and principles is wise.

Therefore, combined knowledge and understanding among all human beings for conservation of biodiversity, exemplified in both formal science as well as local systems of knowledge is the only way towards building a sustainable life.

*"Exclusive truth claims – assertion of epistemological privilege – are now not tenable either on the part of science or local knowledge systems."

47. Counter View

After having written elaborately on managing carbon to satisfy the environmental mindset of planners and historical findings contained in reports, I also thought of writing a reverse piece of glossing the pages of history. From the book of Divine Comedy of Dante, we get to know about the perpetual rainfall cycle between 1300-1400 that shook Europe through crop failures, food shortages, migration and decimation of population. There was no global warming but surely it was a climate change. The North Sea turning icy despite brackish waters much before the dawn of Industrial Revolution just indicates also the irrelevance of modern debate about warming causing climate revulsion.

The history of climate change is not captured in this modern campaign since climate was changing always, decade to decade, region to region without giving notice. It could be due to factors other than CO_2 or acts of providence by itself.

This century is replete with debates through Rio Convention Kyoto Protocol etc. to reduce CO_2 emission levels by allocating tasks and goals and reflecting accountability on nations to find out ways to stop. We speak of abatement technologies of today to reduce pollution but never strike at its roots due to counter demands for growth. We have in the process, overlooked NO_x, SO_x phenomenon affecting population by shifting our focus from fossil fuel to oil and gas as if they are blessed goods for mankind.

The recent research spree is aimed at opening doors for non carbon sources of technology and indulges in modifications in lightening, combusting systems and highlighting solar and electrical energy systems to meet energy appetite of the growing nations.

Probably carbon causing change also existed then but no research was patronized. It was the belief system that governed the Research like Copernicus facing a crusade by saying that Sun was at center of solar system while the Roman Catholic Church had posited Earth at the center of the system.

The thesis to cleanup CO_2 should be confined to pollution only but not to measure temperature deviations causing global warming leading to climate change. It is not tested like the laws of gravitation or the cell phone working through radio waves from transmitters via satellite that travel thousands of kilometers in a matter of seconds.

Trees are more vital in the process of cooling if we believe in warming since they have a greater radiation surface allowing cooling. This aspect is also important to strike at CO_2 if latter is the villain and so massive plantation can be a great tool to make earth safe. The only rational conclusion one can draw is that rather than getting obsessed with science and research to prove things uncertain as certain, it is more meaningful to cut CO_2 emissions that pollute the air since we need to contain it at source to ensure a healthier ecosystem for human beings.

Bibliography

Aam (*Mangifera indica*): 21
Agni: 58, 60, 61, 62, 63, 72
Agro Ecosystems: 14
Ahimsa: 53, 70
Akasha: 58, 59, 60
Algae: 08, 77
Allium stracheyi: 41
Ancient Text: 17, 20, 34, 41, 57
Angelica glauca: 41
Antarctica: 7
Antaryaga: 72
Apas: 58, 60, 63
Arjun (*Terminalia arjuna*): 21
Arthashastra: 20
Asceticism: 54
Asia: 08, 16, 23, 24
Atharva veda: 18, 19
Ayurveda: 72, 73, 93
Bakul (*Mimusops elengi*): 21
Balinese Water Temple: 24
Bamboo drip irrigation: 30
Bawdi: 29
Bengal communities: 42
Bhagavad Gita: 50
Bhils: 51
Bhim: 42
Bhotiya community: 41
Bhudevi: 64
Bhumi: 64
Bio indicators: 77
Biocultural restoration: 13
Biodegradable: 63, 74, 75, 76
Biodiversity Conservation: 13, 14, 15, 32, 33, 34, 43
Bisnoi community: 38
Bolivian Amazon: 24
Brahatsamhita: 21
Brahman: 72
Carbon dioxide: 02, 05, 07, 08, 61, 74, 75

Carbon footprint: 09, 68, 71
Carpool: 86
Chernobyl: 08
Chipko movement: 49, 52
Climate Change: 07, 08, 32, 57, 61, 63, 68, 79, 90, 94, 95
CNG: 92
Compact Fluorescent Light (CFLs): 07, 89
Conservation principles: 17, 18
Convention on Biological Diversity: 91
Cultural and religious values: 46
Dang district: 45
Dehi me dadami te: 46
Desertification: 35
Developing Countries: 07
Dharani: 64
Dharma: 49, 52, 53, 56, 70
Durga: 43
Eco-friendly: 81
Ecology: 09, 16, 18, 31, 46, 51
Elephant forests: 336
Emissions: 07, 69, 71
Endangered: 07, 42
Energy Conservation: 89
Environmental Education: 46, 47, 48
EP act: 82
Equity of knowledge: 90
Ethical values: 47
Ethnoforestry: 23, 33
Farm biodiversity: 40
Farming strategies: 14
Forest management: 32, 33
Forestry policy: 22
Formal education: 48
Formal science: 31, 90, 91, 93
Gandhi: 49, 54
Ganesh idol: 81
Ganges: 50
Ganpati festival: 81, 82

www.ingramcontent.com/pod-product-compliance
Lightning Source LLC
Chambersburg PA
CBHW050518190326
41458CB00005B/1587